D1642265

Mathematical Sorcery

Revealing the Secrets of Numbers

Mathematical Sorcery
Revealing the Secrets of Numbers

Calvin C. Clawson

PLENUM TRADE • NEW YORK AND LONDON

Library of Congress Cataloging-in-Publication Data

Clawson, Calvin C.
Mathematical sorcery : revealing the secrets of numbers / Calvin C. Clawson.
p. cm.
Includes bibliographical references and index.
ISBN 0-306-46003-3
1. Mathematics--Popular works. I. Title.
QA93.C62 1999
510--dc21 99-13848
CIP

ISBN 0-306-46003-3

Plenum Trade is a Division of Plenum Publishing Corporation
233 Spring Street, New York, N.Y. 10013

10 9 8 7 6 5 4 3 2 1

A C.I.P. record for this book is available from the Library of Congress

Printed in the United States of America

For
Sue and Zyg
Ann and Cal

Contents

Acknowledgments ix

Introduction 1

Chapter 1: Early Counting 9

Chapter 2: The Incredible Greeks 22

Chapter 3: Mathematical Proofs 55

Chapter 4: Passing the Torch 80

Chapter 5: Opening the Door 115

Chapter 6: Functions 133

Chapter 7: Stretching Space 176

Chapter 8: Extending the Form 211

Chapter 9: Isaac Newton 228

Chapter 10: Calculus 248

Chapter 11: Speculations on the Nature of Mathematics 281

Endnotes 286

Index 291

Acknowledgments

I wish to acknowledge my indebtedness to my wonderful workshop friends who were so patient in their helpful review of my efforts: Marie Edwards, Bruce Taylor, Linda Shepherd, Phyllis Lambert, and Brian Herbert. I want to thank Robert Casad, Dave Wetzel, and Larry Curnutt for their courage in plowing through the manuscript to weed out the errors and corruptions.

A very special thanks goes to my editor, Linda Greenspan Regan, whose continuing support and encouragement made the entire project possible. Without her constant care and guardianship this book could never have been completed.

Introduction

In the classes I teach at a local community college, I always begin the first lecture by asking my students why they are enrolled in a math class, writing the reasons they suggest on the blackboard. Their answers are always sound and, in themselves, sufficient justification to take a college course in mathematics. Some of their reasons include: I need this course for graduation; math will help me in my future job; it can help me manage my personal finances; and it will prepare me to teach my children. Yet, all of these reasons miss the mark.

I then inform the class that I will write the primary reason for our participation in a mathematics class on the board. I write in large letters "FUN." They laugh. I know what they are thinking: "What planet did this screwball step off of?" But no one would react this way if I were teaching art or music. Mathematics, like art and music, has the potential to add great depth and meaning to our lives, and to enrich each of our days, long into our twilight years. In truth, a world without mathematics would be a dull place, indeed. However, the vast majority of my young students have a real distaste for mathematics. In most cases, they are taking my class because they must—they are required to have a specific level of mathematical competence to graduate. If math courses were not required, I would be out of work. How did mathematics get such a bad reputation?

The primary reason our children grow into adulthood without any affection for mathematics is because of how we teach it. Mathematics is a required subject in elementary school. Every elementary teacher must teach his or her students the minimal amount. For a good number of teachers, however, math is not their favorite subject. For most of them, it was not even one of their preferred subjects while in college. As teachers, they cover the required subject matter, but without real enthusiasm. The students are proficient at reading their teacher's body language, and they quickly learn that mathematics is not their instructor's favorite subject. So why should it be theirs? After all, the teacher is the adult in the classroom. If the adult doesn't like mathematics, there must be something wrong with it. This perception is seldom a problem when our children learn art or music, for they are taught by teachers who teach them with great zeal.

Mathematics is as much an art form as is art and music. It deserves to be taught by teachers who love it, just as music teachers love music and art teachers love art. Unfortunately, it doesn't appear that the current situation is about to change. We continue to ask our children to learn mathematics from teachers, many of whom, themselves, are bored and tired—even afraid—of it. Some students are fortunate enough to encounter good math teachers in high school, others must wait until college. Some students never have the chance to learn from a teacher devoted to mathematics.

Adults fan the fears of mathematics by perpetrating damaging myths that have sprung up this century. Most adult Americans feel no affinity for mathematics, which has both spawned these myths and then promoted their acceptance within the general population. That we live with these fictions is both a disservice to mathematics and a denial of our full potential as modern human beings. In addition, the mathematical abilities of American children now lag far behind those of children in other technological societies, which puts our young people at a significant disadvantage when competing in the world's marketplaces.

In short, these myths, simply stated, are:

1. Mathematics is boring.
2. Mathematicians are dull old men with beards, hidden away on college campuses, who have lost touch with reality.
3. There are two kinds of people: those who can do mathematics, and all the rest of us.
4. Girls (women) have no facility for mathematics (and really don't need it, anyway).

I have excellent news: these myths are all false. Yet, most adults believe most of them. Let's consider the first statement: mathematics is boring. Nothing could be further from the truth. Mathematics is not only interesting, it is *electrifying*, it is *astounding*, it is the *alpha and omega*.

Now some of you may be laughing and thinking that I'm just throwing words around to generate hype. Yet, silently slip into any group of mathematicians and listen to their conversation. They will be talking lovingly about mathematics. It's not just puppy love, but a full blown, absorbing passion. How can this be? How can the majority of Americans dislike something mathematicians find so fulfilling? That mystery, in part, is what this book is about.

We move now to our second myth: mathematicians are dull old men with beards, hidden away on college campuses, who have lost touch with reality. In the secret recesses of your heart, haven't you harbored this very idea? Yet, such people represent only a small percentage of the world's mathematicians, for there are many mathematicians scattered around the globe who find their homes in industry and business. If we broadly define a mathematician as one who receives his or her primary income from doing or teaching mathematics, then every mathematics teacher is a mathematician.

On occasion I receive letters from readers, generally people who do not consider themselves professional mathematicians. One such letter came from Leonard Chmiel of Salem, Massachusetts, a 55-year-old retired businessman with a high school education. Mr. Chmiel presented me with two mathematical ideas. The first was a complex presentation about prime stars, which, frankly, I never did figure out. But the second was a test for prime numbers.

During the last two decades, mathematicians have focused considerable attention on testing large numbers to discover which are prime and which are not. A prime number is a positive integer that can only be evenly divided by 1 and itself. Two examples are 7 and 11. All other numbers (excluding 1), such as 6, are called composite numbers. Modern encryption codes employed by business and industry utilize large composite numbers (which are the product of two large prime numbers) to encode messages sent by computers. Hence, cryptologists are interested in both testing large numbers for primeness and learning new ways to factor large composite numbers into primes.

Therefore, when I saw Leonard's test for primes I was naturally curious. I tried his test on several numbers and it worked every time! Had Leonard discovered a new test for prime numbers? I decided to try writing Leonard's procedure in the symbolic language of algebra. After a few manipulations I finally got an equation that expressed the same idea as Leonard's algorithm.[1] The equation looked familiar, so I dug out one of my number theory books. I discovered that Leonard's algorithm is one instance of a famous theorem discovered by Pierre de Fermat (1601–1665), the father of modern number theory. He achieved this approximately 350 years ago, and it is called Fermat's Little Theorem. I have never stumbled upon such an important theorem while playing with mathematics, nor, to my knowledge, have any of my professional colleagues. Yet, Leonard did. If he had done this 350 years ago, the theorem might now be called "Leonard's Little Theorem."

Another example of a mathematician outside the ivy tower tradition would be the case of my college friend, Randy.[2] Randy was raised in Nevada and worked as a cowboy for most of his adult life. Never having completed high school, he was not much enchanted with books or learning. Then one day he killed a man in a fight over a woman. He was sentenced to the Utah State Penitentiary for manslaughter. While in prison he began to read books to battle the boredom. One day he picked up a mathematics book. From that day forward he was intrigued with numbers and mathematical relations and taught himself the basics of mathematics while still in prison. After being paroled, he took the entrance examination for the University of Utah. His score was so high that the university overlooked his lack of a high school diploma. I met Randy at the university, and we took several classes together. He was still very much a cowboy in word and dress. Yet, he was a mathematician, too. After we graduated, Randy took a job as a mathematician for a Canadian company.

From the above examples we should learn a valuable lesson. Mathematics is not the private domain of professors at prestigious universities. It is a subject enjoyed by countless individuals everywhere. Mathematics can be carried out by anyone who is interested in the subject and willing to experiment by playing with numbers and their relationships. Pierre de Fermat was an amateur mathematician, yet he cofounded analytical geometry, pioneered in the field of calculus, and established the modern subject of number theory. Therefore we see the need to adopt our broader definition of just what a mathematician is.

A mathematician is anyone who either receives the major portion of his or her salary from doing or teaching mathematics, or one who both loves mathematics and performs mathematics for the sheer joy of it. Martin Gardner, who wrote the "Mathematical Games" column for *Scientific American* for almost twenty-five years, is the most prolific writer of mathematical books in the twentieth century, yet he is not a professionally trained mathematician. However, I would venture to say he knows more fundamental mathematics than just about anyone alive today.

We are now ready to smash myth three: there are two kinds of people: those who can do mathematics and all the rest of us. I can't recall how many times I've heard this statement. My own mother used to say the same thing, "I just don't have a mathematical mind. If someone says 'the train left the station at such and such a time. . .' well, I just know I'm lost and I stop right there." Many people believe that babies are born with either brains that are good for mathematics or brains which are not. It doesn't matter what the no-math brained babies do during their lives, they will always be poor at mathematics. This thesis is nonsense.

It is true that some people seem to be able to grasp abstract relationships quickly. These people learn mathematics rapidly. The biographies of famous mathematicians generally demonstrate that they possessed such talents early in life. However, some people want to generalize this fact to the idea of two types of brains, which then becomes their excuse not to study mathematics. If you have a no-math brain, why should you bother to struggle with the subject?

We should also realize that many different talents come to people in different degrees and at different times in their lives. Because some people have a "musical ear" and learn to play music effortlessly does not mean that the rest of us should not learn music. Or, that because Tiger Woods is a highly gifted golfer, all the rest of us golfers should just put away our clubs. The truth is, the vast majority of people (including you and me) fall into the middle of the bell curve of intellectual talent. We can, if we desire, learn all the mathematics we want. When you see an individual performing complex mathematical manipulations with ease, you're tempted to assume he or she possesses one of those fictitious math-brains. However, the ability demonstrated is one achieved through long years of practice. Mathematics is much closer to athletics and music than people realize. If you practice at mathematics you become good at it. My hope is to put to rest the myth of math-brains.

Myth four: girls (women) aren't good at mathematics and, its corollary, they really don't need it, anyway. As a society we have finally come to realize this idea is wrong. Yet, in many situations we still act as if it were true. Sixth grade girls, on average, perform above their male counterparts on standardized math tests. However, by the twelfth grade, the situation has reversed, with the boys outscoring the girls.

What happened in those years between sixth and twelfth grade? Teachers, parents, and society at large send the message to girls that they are not good in mathematics, that they really don't need math, and "anyway you'd better not embarrass your boyfriend by doing better." Not only is this attitude terribly unfair to the millions of young women who would both enjoy and make contributions to the field of mathematics, but it is damaging to our society as a whole. We need more mathematicians, both as teachers and mathematical innovators in industry. Yet, with our biased attitudes we eliminate half of our young people from discovering a career in mathematics. How many young virtuoso mathematicians have we lost in this way?

It is not sufficient that we simply dismiss the four myths we have discussed; we must go further. We must change the way society defines what a good education is. In certain past societies, mathematics played a role in a well-rounded education. We see this in ancient Greece, when an education included not only music, rhetoric, and science, but mathematics in particular. Above Plato's Academy, one of the first and greatest schools the world has ever known, were inscribed the words, "Let no one ignorant of geometry enter here."

Our colleges graduate great numbers of liberal arts and science students, many who have no grasp of what mathematics really is. Instead, they possess, at best, a limited set of calculating procedures meant to grind out answers to specialized problems. When encouraged to take more math, thus broadening their mathematical backgrounds, they say, "what will I ever use it for?" Mathematics is the language of science, and is a great utility in industry and business. Hence, mathematics gives the student not only great power to solve difficult real-world problems, but helps the student to understand how the universe operates. We must include mathematics not only for the power it bestows upon us for problem solving, but because mathematics enhances our understanding of the most basic processes we encounter as inhabitants of this universe. These mathematical ideas then become a goal unto themselves, for they are a source of direct and immediate pleasure. It is our intent to present within these pages a limited number of these ideas so that you, the reader, can experience the mathematician's delight.

For this romp through mathematics you will need a reasonable ability to do basic arithmetic computations and a basic understanding of algebra. In some places I will pause to remind you of your algebra training. With this background you will come to understand the great ideas of modern mathematics, including analytical geometry, calculus, and matrix algebra. It is going to be great fun. Truly.

Before proceeding let us state a thesis that we will be considering as we watch mathematics explode from the simple arithmetic of the ancient Sumerians to a wonderful complex world as we approach the twenty-first century. This thesis is that much of mathematics grows from a combination of three intellectual forces: (1) generalization, (2) extension into the infinite, and (3) improved symbolism.

The act of generalization is when we expand a definition to include more objects or areas of thought. It may also occur when we have several disjointed areas of study that we can combine into one conceptualized area. As we study the history of mathematics we will see how the need to find meaningful solutions to problems forced mathematicians to expand the set of objects defined as mathematical, and that this resulted in much more powerful mathematical systems. This process is not unlike our generalization of the concept of being human. Many primitive tribes thought of themselves as the only humans, while they considered everyone else to be barbarians or nonhumans. Since those times we have expanded the idea of humanity to include all *Homo sapiens sapiens*.

The second force, number extension, occurs when we are willing to expand our numbering system into the infinite. A battle raged within mathematics between those who would limit our thought to only finite sets of objects and those who were anxious to jump into infinite collections. This debate has not been completely settled today.

The last force is the need to expand our notational system to include simpler forms in some situations and more complex forms in others. Hence, the growth of mathematics was either slowed or advanced by how quickly mathematicians could invent symbolic algebra. Just as written language acquired improved notation with the adoption of vowels and consonants, mathematics improved its ability to serve by inventing new symbolism. Therefore we want to watch this triad: generalization, number extension, and symbolism, to see how they allowed mathematics to evolve.

Finally, we will end our journey by asking a very philosophical question: what difference does mathematics make to the universe at large? Is mathematics just a complex game invented by humans that will disappear completely when we are all gone, or is it part of the very script that writes the universe, and us, into existence?

I hope the following pages will provide you with some insight into the exquisite beauty of mathematics, so you can feel the rapture that mathematicians feel. As with both music and fine art, we must learn to see what is within the mathematical subject that elicits such emotion from its practitioners. In preaching a love of mathematics I am at a disadvantage if compared to the lover of music or art. Both of these disciplines present the audience with a complete product that is immediately pleasing to the senses. In music and art, the eyes and ears respond to the sights and sounds of intricate beauty. However, much of mathematics is not a subject that is directed toward pleasing any of the physical senses, be it vision or hearing. Rather, mathematics is an enchanting diversion of the intellect. The intricate pictures and drawings we may employ to convey a message are not to be confused with mathematics itself. To appreciate mathematics requires us to arouse our intellectual faculties, not our senses.

To join with me in mathematical rapture, you must become a participant and construct within your own mind the wondrous relationships I present. Mathematics cannot be enjoyed passively, as art and music can be, but requires your active involvement. I invite you to come with me and take part as a partner in the great intellectual adventure we are about to begin, and discover in the deepest way how mathematics excites such passions in the human soul.

I

Early Counting

Mathematics began with counting and numbers. To understand the roots of mathematics, then, we must go back to the time when numbers were first used—a most difficult task. Yet, such a journey will help us discover whether mathematics is a secondary characteristic of human beings, something we accidently invented, or whether it is really a basic part of our nature. If mathematics is a fundamental part of being human, then it strengthens our belief that mathematics can and should be enjoyed by a broad spectrum of our society, and not just a few "bearded old men."

Understanding the origins of counting will facilitate our understanding of the nature of mathematics and help us identify those forces which have driven its creation. The first motivation for people to create numbers was the human desire to know the manyness of a set of objects. In other words, to know how many duck eggs are to be divided amongst family members or even how many days until the tribe reaches the next watering hole. How many days will it be until the days grow longer and the nights shorter, how many arrowheads do I trade for that canoe? Knowing how to determine the manyness of a collection of objects must surely have been a great aid in all areas of human endeavor.

To achieve this required the act of counting, which mathematicians refer to as "one-to-one mapping." When we count on our fingers, we assign (map) one unique number word to one specific finger. Hence, when we are done, each of the first ten number words are assigned to one of ten different fingers. The order in which we do this is important, for it guarantees that the last number word uttered is the count of the set of objects counted, i.e., the manyness of the set. The count of a set of objects is its measure of manyness.

This same mapping is carried out when we physically count any set or collection of objects. We point with our finger and say the appropriate number word in the appropriate sequence. When done, the last number word spoken is the count we are after. From this wonderful counting process we get the counting numbers: 1, 2, 3, 4, 5, . . . The three dots (. . .) after 5 is an ellipsis, and indicates that, potentially, the count can go on indefinitely.

The earliest direct evidence of counting are two animal bones which show clear grouped marks. One is a 35,000-year-old baboon's thigh bone from the Lebembo Mountains of Africa, and the other is a 33,000-year-old wolf bone from Czechoslovakia (Figure 1). The wolf bone, found at an ancient human campsite, is especially intriguing. It was notched with fifty-five marks, grouped in eleven sets of five marks each. While 33,000 years is long before the beginning of farming (10,000 years ago) and the growth of cities, it was still a time of our immediate ancestors—*Homo sapiens sapiens* or modern humans. These two bones demonstrate that our early hunter–gatherer ancestors were counting.

Could counting be even older? Those humanoids living in Europe and the Middle East before modern humans, going back as far as 130,000 years, were the Neanderthals. They were definitely not modern humans, but did have brains larger than *Homo sapiens*. The Neanderthal average adult brain was approximately 1500 cm^3 (3.3 lbs.) compared to our average size of 1300 cm^3 (2.9 lbs.). We have no direct evidence that they counted, but we have tantalizing clues that they may have. They were intelligent enough to build shelters, use fire, make sophisticated tools, and bury their dead with flowers. They probably participated in religious rituals.

In April 1996 Oscar Todkoph of Hindenburg University discovered a

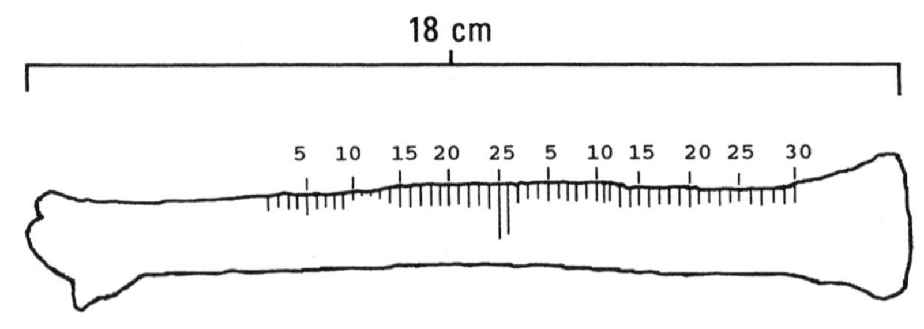

FIGURE 1. Drawing of a 33,000-year-old wolf bone found in Czechoslovakia in 1937 by Dr. Karl Absolon. The fifty-five grouped notches demonstrate that counting was a talent possessed by its maker. Drawing adapted from photograph, Lucus Bunt, Phillip Jones, Jack Bedient, *The Historical Roots of Elementary Mathematics* (New York: Dover Publications, p. 2).

50,000-year-old mastodon tusk which had sixteen aligned holes in the surface.[1] He believes it was a musical instrument. This artifact predates the appearance of modern humans in Europe, which makes the artisan a Neanderthal. At another Neanderthal site archaeologists found a hollow bone with holes deliberately drilled through to the core. It may have been a Neanderthal flute. These bones appear to demonstrate that the Neanderthals participated in music—a very human characteristic. Could they count? We don't know, so we must wait to see if some lucky archaeologist discovers direct evidence.

If the Neanderthals did count, it could push back counting 130,000 years. Is it possible that counting is older? A hominoid considerably older than Neanderthal was *Homo erectus*, who flourished from 1.5 million years ago until approximately 300,000 years ago. *Homo erectus* may have been our early ancestor. They did not have the brain power of either modern humans or the Neanderthal, for their brain cases ranged from 900 cm^3 (2 lbs.) to 1100 cm^3 (2.4 lbs.) or approximately midway between modern chimpanzees (450 cm^3 or 1 lb) and humans. Yet, they were remarkably intelligent if we consider their varied activities. They domesticated fire and immigrated from Africa to Asia and Europe. They made sophisticated tools and built shelters. Did they count? Again, we don't know, and we place ourselves in the hands of the archaeologists for more evidence. It is not wild speculation to believe that archaeological evidence might appear to push back the introduction of counting. New discoveries are constantly forcing us to revise upward our estimate of the intelligence and craftiness of our ancestors. Many depictions of *Homo erectus* show a stooped, hairy brute pounding the ground with a club or rock. However, in 1994, Hartmut Thieme of the Institute for Historical Preservation in Hannover, Germany discovered a cache of 400,000-year-old spears in a coal pit east of Hannover.[2] These well-crafted, weighted spears were probably made by a late *Homo erectus*. Their presence demonstrates that these people were not mindless brutes, but probably skillful and talented hunters.

Anatomical evidence exists that suggests counting is a very ancient activity. Neurologists have determined that certain activities are associated with specific parts of the brain. Hence, language is generally associated with the left side of the brain, which also is the home for functions such as planning and executing sequential operations. One specific area of the left rear of the brain is the Gerstmann area, which gives us the ability to both recognize our fingers and carry out simple arithmetic calculations.

That we associate finger recognition and calculating skills should not be surprising, since we first learn to count as children by associating fingers with numbers. The fact that both activities are assigned to the same part of our brain suggests we have been counting on our fingers for a long time.

If counting stretches back many hundreds of thousands of years and is in some way hardwired into our brains, then counting and numbers are part of our very natures. To be human is to count and know numbers. The idea is reinforced if we observe how much entertainment people derive from numbers, even those who generally profess a disdain for mathematics. Many of our games use numbers. Blackjack is, after all, a counting game; can I add my cards to get closer to 21 than the dealer, without going over? Craps is also a numbers game, and craps players become adapt at calculating rather complex odds on their bets. You may have been to a horse race and watched handicappers frantically computing indices to identify the winner of the next race. We incorporate numbers into our music. We use them to identify our houses and phone numbers. People study all kinds of complex number indices to watch the stock market. The use of numbers, counting, and simple arithmetic is everywhere. If humans can be described as the toolmaking or fire-using ape, then another appropriate description for us is the counting ape.

ABSTRACTING NUMBERS

Around 10,000 years ago (8000 B.C.), something very dynamic happened to humans on this planet. In an area known as the Fertile Crescent, including parts of Israel, Southern Turkey, and the Tigris–Euphrates Valley of Iraq, people put down their hunting implements and began farming. The rise of farming enabled towns to grow, for people could now live in the same place year round. There was no need for them to migrate in pursuit of herds of game animals, as was the custom of their cousins, the hunter–gatherers. However, they did have to perform additional sophisticated tasks, which were alien to their hunting cousins. They had to plan how much seed to plant and how much to save. They had to protect their fields, store their harvest, ration food. Once towns grew, they had to protect themselves from marauders who would literally steal the fruit of their labors. In order to organize soldiers and build fortifications, government was born. To pay for keeping the soldiers and building the forts, government needed taxes!

All these innovations put pressure on our ancestors' mathematical abilities, for it was no longer sufficient to just count things. It became necessary to add, divide, subtract, and multiply; the four basic operations of arithmetic now became indispensable. What evidence do we have of arithmetic for this time. No written evidence exists since writing wasn't invented until approximately 3100 B.C. This shortage of records leaves us to wonder what occurred during the 5000 years between 8000 B.C. and the beginning of recorded history.

For many years anthropologists found funny little clay objects in the ancient settlements of Western Asia, the birthplace of early farming. How were these objects used? Some authorities suggested they may have been a form of primitive chess, or possibly fertility figures. But tens of thousands were found, many in sites that were once ordinary homes. Were that many games of prototype chess played back then? In the 1970s Denise Schmandt–Besserat of the University of Texas discovered the answer. After cataloging over 8000 of these objects taken from 116 sites in Iran, Iraq, and Turkey, she determined that they were counting tokens.[3] The farmers and other villagers were using small baked clay objects to keep track of their belongings. Some of these counting tokens were wrapped in soft clay which was then baked in hearths, creating a hard envelope and providing a kind of sealed record of a count. While certain token shapes indicated a specific number, other tokens represented specific objects counted—be they sheep or jars of wine.

With the tokens safely sealed inside it, the envelope could be transported with a shipment of goods to act as an official tally of those goods. Yet, at times, it was necessary to know how many tokens were inside the envelope without actually breaking it open. This would occur when goods were being transferred from one shipper to another. Thus users developed the habit of impressing the shape of the tokens which were on the inside of the envelope onto the outside of the envelope before it was baked. When the baking was completed, one could "read" which tokens were inside based on the imprints on the outside, and, if necessary, break open the envelope to verify the count with the actual tokens.

After several thousands of years, some smart individual realized that all this work was unnecessary to transmit a simple count. Why not just take a small soft clay tablet, impress the shape of the tokens on the surface, and then bake the tablet? This yielded the oldest tablets from ancient Sumer, the birthplace of writing in Western Asia. Sumer was a kingdom occupying the southern half of modern Iraq and included the ancient cities of Ur, Larsa, Uruk, and Eridu. The control of this rich farming plain containing the Tigris and Euphrates Rivers shifted almost constantly between these cities. Yet it was on this plain that written history first appeared.

For many centuries scholars were uncertain as to how writing began. Some speculated that it grew from religious ritual, or that the priests invented it in their spare time. Many myths claim the gods, or God, gave humans writing as a gift. We now know that the first writing in Western Asia grew out of a need to record and track financial transactions. The oldest clay tablets from Sumerian cities prior to 3100 B.C. show the imprints of number tokens, representing various counts, plus drawings or pictograms of commodities, such as cattle or bushels of grain. This was the first prototype of writing. It is significant that the birth of writing was so closely connected to the need to record counts and the use of numbers. Soon, symbols for proper names appeared (to identify the owner of the goods) followed by the invention of verbs to show actions. By 3100 the Sumerians were using a full-fledged written language to record their history, which they wrote as large clay cuneiform texts. In conjunction with the writing came the expansion of mathematics.

Strong evidence exists that numbers, when first used, were not generalized as objects themselves, but treated as adjectives describing different sets. In many primitive languages, different number words existed for different objects. We still see remnants of this system in English when we speak of a brace of oxen and a pair of gloves. In primitive societies, different number words were frequently used for different objects. For example, Fiji Island natives use the word "bola" for ten boats, but the word "koro" for ten coconuts.[4] Numbers were not considered as things in themselves, but as attributes of sets of concrete objects. Therefore numbers could not exist without some collection of concrete objects which they counted. Denise Schmandt–Besserat divides the development of counting into three stages: one-to-one counting, which we do when we count on our fingers; concrete counting, when we assign a unique number word to a specific collection of objects; and abstract counting, when we allow the numbers to stand by themselves as independent mental objects. She then goes on to say:

> . . . in the Near East abstract counting may have been preceded by an archaic concrete counting system, using different numerations to count different items. The third-millennium cuneiform texts leave no doubt of

> the fact that the Sumerians had developed a most elaborate sexagesimal system of counting. There is also no doubt that their arithmetic was based on abstract counting.[5]

The earliest tokens were used to represent counts for specific commodities. Ovoid-shaped tokens counted jars of oil, while spheres counted grain. This is similar to the use of specific group names such as a yoke of oxen and a pair of gloves. It would have been as wrong for an early Sumerian to count jars of oil with spheres as it would for us to talk of a yoke of gloves. Assigning specific number symbols to particular objects is concrete counting, for the numbers always refer to concrete things. When we shift to abstract counting, we use only one set of number words or symbols which can apply to any group or can be used to describe relations amongst the symbols (numerals). Schmandt–Besserat even credits this advancement from concrete numbers to abstract numbers around 3100 B.C. to a specific city: Uruk, which was just northwest of the confluence of the Tigris and Euphrates Rivers in southern Iraq.[6]

The emergence of the Sumerian numbering system, beginning with the use of clay counting tokens and ending with a sophisticated writing system and abstract numbers, represents two of the three generating forces that propel mathematics. First, we have the generalization or abstraction of numbers. For instance, numbers were no longer simply attributes ascribed to specific physical objects, but began to acquire an independent existence of their own. Only one number, "two," was needed to represent any collection of two objects, and we no longer needed different names for different sets containing two elements. This does not appear to have happened all at once, but to have evolved over many years. This may sound like a trivial accomplishment, yet it is a magnificent leap of intellectual prowess. Once the farmers of ancient Western Asia abandoned the need to have a new set of symbols and names for every different set of concrete objects, and used only one set of symbols, they greatly increased their power to model and comprehend the world they lived in. If we add two coconuts to three coconuts, we get five coconuts. If we do the same addition with bananas, do we get five bananas? We know we do because we also know that for any discrete collection, $2 + 3 = 5$. Therefore, by generalizing numbers we greatly expand their power. The ability to generalize what we call numbers has driven mathematics into many new and exciting areas as we will see in the coming pages.

The second great motivating force having an impact on mathematics was the development of a symbolism to allow us to record and manipulate numbers. This began in Sumer when the Sumerians started to impress the shape of their tokens into soft clay tablets before baking them. Hence, numbers could now be represented by written symbols. At first, this process was a one-to-one concrete symbolism, but eventually, symbols evolved that stood for idealized collections of objects. The written number systems which evolved for the Sumerians, and later for the Babylonians who conquered the Sumerians, are shown in Figure 2. Notice that in both the Sumerian and Babylonian systems the symbols for the count of 1 through 9 are on a one-to-one basis. That is, each time we add one to the count, we simply add one more symbol. Finally, when this becomes too cumbersome, this one-to-one method is abandoned and we substitute one symbol for a count of 10. Again we go to a one-to-one system until we reach 60, where a new symbol is introduced. Significantly, these systems not only use numerals, but also use numbers which no longer represent specific objects, numbers that can stand for any discrete collection of objects. This means that if we can manipulate these symbols to discover a relationship between them, this relationship will be true of apples, bananas, houses, horses, etc.

There exists a third driving force to expand mathematics which we do not encounter with the Sumerians. This third force is to extend the count from a finite set to an infinite set. While the Sumerians had some notion of the infinite, we do not see it in their mathematics.

THE WONDERFUL EGYPTIANS

In the ancient world there stood no greater empire than that of the Egyptians. In longevity alone it beat all other ancient kingdoms. For many thousands of years people lived along the Nile River in northern Africa, and by the Copper Age (4000 B.C.) began establishing permanent farming settlements. These settlements grew and separated into Upper Egypt and Lower Egypt. In 3100 B.C., Menes, the king of Upper Egypt, united the two kingdoms and the great line of pharaohs began. By 2900 they were beginning to build pyramids, and by 2560 Khufu was building his Great Pyramid. Their kingdom continued, almost uninterrupted, until 525 B.C., when Cambyses II, son of Cyrus the Great, King of Persia, defeated the last Egyptian pharaoh. Thus ended a mighty empire that had lasted for approximately the first 2600 years of humankind's recorded history.

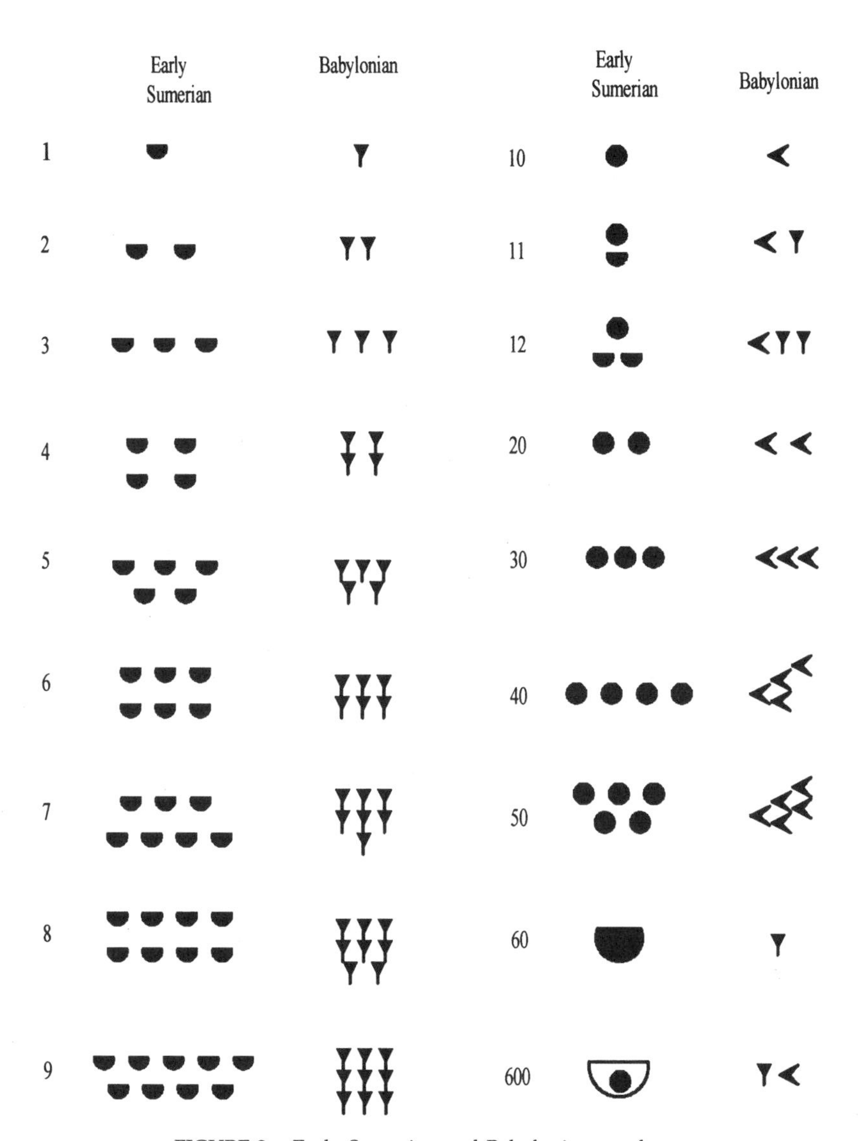

FIGURE 2. Early Sumerian and Babylonian numbers.

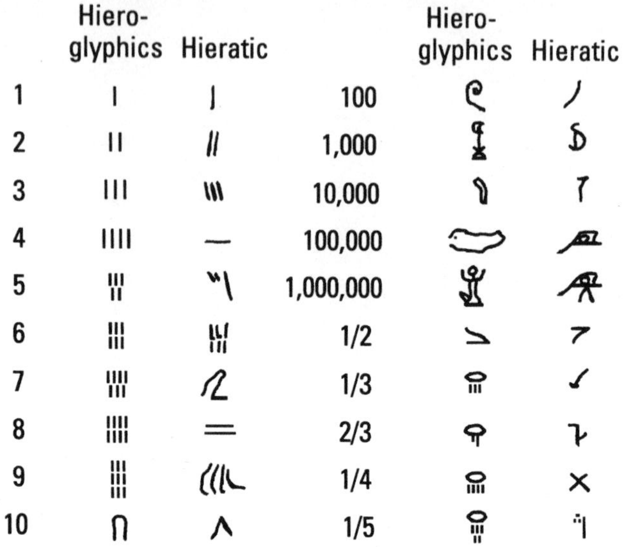

	Hieroglyphics	Hieratic		Hieroglyphics	Hieratic
1	I	/	100	𓍢	/
2	II	//	1,000	𓆼	[illegible]
3	III	\\\	10,000	𓂭	[illegible]
4	IIII	—	100,000	𓆐	[illegible]
5	III II	[illegible]	1,000,000	𓁨	[illegible]
6	III III	[illegible]	1/2	[illegible]	7
7	IIII III	[illegible]	1/3	[illegible]	[illegible]
8	IIII IIII	=	2/3	[illegible]	[illegible]
9	III III III	[illegible]	1/4	[illegible]	×
10	∩	∧	1/5	[illegible]	[illegible]

FIGURE 3. Ancient Egyptian hieroglyphics and hieratic numerals.

The appearance of writing in Egypt was almost contemporary with that of Sumer, yet most authorities consider it to be later, and may have been influenced by the Sumerian writing. The Egyptians evolved two sets of symbols for their numbers (Figure 3). One set was in the official written language of hieroglyphics, used on monuments and public buildings, while the other set was in the hieratic written language, used for day-to-day commerce. Notice that the beginning numbers from 1 through 9 in hieroglyphics show the one-to-one counting characteristic. A new symbol is then introduced for 10, and then additional ones are added up to and including the number for 1 million. We can assume from their accomplishments in construction, including their pyramids and temples, that the Egyptians developed their mathematics during the first half of the third millennium B.C. However, we have discovered no mathematical documents earlier than the beginning of the second millennium, many hundreds of years later. The oldest mathematical text is the Moscow papyrus which dates from approximately 1890 B.C. A second famous papyrus written around 1650 B.C. may have been copied from an older one dated as early as 2000 to 1800 B.C. That the Egyptians were carrying out sophisticated arithmetic computations many hundreds of years before this is certain, for the sides of the Great Pyramid are accurate to within 2/3 of an inch (1/14,000 of the total length) and the angles at the corners of the Great Pyramid are at most in error by only 12 seconds of a degree (1/27,000 of 90 degrees).[7]

We know that the Egyptians used fractions, but they were only unit fractions, where the numerator is a 1 while the denominator can be any positive whole number. For example, to represent the fraction $3/7$ the Egyptians wrote it as the sum of the unit fractions, $1/3 + 1/11 + 1/231$. We see at once the great awkwardness of this system for it requires us to complete cumbersome calculations just to add or subtract simple fractions. To make their work easier, the Egyptians used extensive tables of unit fractions in performing computations on fractions.

The Egyptians didn't evolve a more efficient system for recording fractions because of their system of symbols. To represent a unit fraction they drew a small oval or dot above a whole number that represented the fraction's denominator. This made it easy to write unit fractions but impossible to write a fraction with a numerator greater than 1 unless a special symbol were invented. Hence it was the Egyptian symbolism that prevented them from expanding to a more useful form.

In addition to performing the basic operations of addition, subtraction, multiplication, and division to great accuracy, the Egyptians could solve problems of elementary algebra; compute areas of regular shapes such as triangles and trapezoids; and calculate the volumes of cylinders and pyramids.

THE SUMERIANS AND BABYLONIANS

During the third millennium, while the Egyptians were codifying their mathematics and building their temples and pyramids, the Sumerians were also developing their mathematics. The Sumerian numbering system used a combination of symbols for both 10 and 60, and is known as sexagesimal. By 2400 B.C. they had begun using fractions. With their sexagesimal system they could write both very large numbers and very small numbers. While the Egyptians used unit fractions, the Sumerians used a fraction of the form $n/60$ or $n/60^2$. This was achieved by simply writing the number n, and requiring the reader to determine if the number was supposed to be n or $n/60$. For example, 30 was written as <<<, where each < represented 10. The fraction ½ was also written as <<<, which meant 30/60 or ½. The advantage of the Sumerian system was that fractions could be easily written and manipulated. The drawback was that the reader had to know from the context whether whole numbers or fractions were meant.

The introduction of fractions for both the Egyptians and Sumerians is significant because it marks the first large expansion of the numbering system—from the use of positive whole integers (counting numbers) to whole numbers plus parts of whole numbers (fractions). In symbolic terms we say a fraction is any number written as p/q where both p and q are positive whole numbers. The system of fractions actually includes all positive whole numbers since every whole number, n, can be written in the form of a fraction as $n/1$. This represents an example of the force of generalization mentioned earlier. By generalizing the idea of number to include both positive whole numbers and positive fractions, the ancients greatly increased the power of mathematics to model the physical universe. To count the number of elements in a set, all we need are the whole numbers, but when it comes to dividing something up into parts, then we need the fractions. Hence, in construction, fractions become invaluable in measuring various lengths. In terms of symbolism, the Sumerians surpassed the Egyptians, who were tied down to the unit fraction. The Sumerians could approximate any fraction with their sexagesimal system.

The political power of the Sumerians shifted between different Sumerian cities until 1763 B.C. when Hammurabi, king of Babylon, defeated the reigning Sumerian king, Rim–Sin of Larsa. This ended the Sumerian empire and began the 1200-year history of Babylonia, the second great empire of ancient Western Asia. It was the Persian king, Cyrus the Great, father of Cambyses II who brought down the Egyptian Empire, who defeated the Babylonians, ending their kingdom in 539 B.C.

The Babylonians adopted both the writing and mathematics of the Sumerians, and then added to both. The Babylonians could find the positive roots of quadratic equations, which have the form $ax^2 + bx + c = 0$, where a, b, and c are constants. Quadratic equations contain at least one term where the unknown has been squared. They could also solve certain cubic equations, $ax^3 + bx^2 + cx + d = 0$, but did not have a general solution for all such equations. They used tables to compute compound interest and were familiar with the Pythagorean Theorem, which states that for a right triangle the sum of the squares of the legs equals the square of the hypotenuse. They made reasonably good approximations of the value of $\sqrt{2}$.

The sexagesimal system of the Sumerians and Babylonians can be seen today. They divided the day into twenty-four hours, the hour into sixty minutes, and a minute into sixty seconds. By 2000 B.C. they were using a calendar of 360 days divided into twelve months of thirty days each. To make up for the extra five days per year, they inserted an extra month every six years. They divided the circle into 360 degrees, probably because their year was 360 days.

We must not be fooled into thinking that the Sumerians, Babylonians, and Egyptians were the sole inventors of mathematics. Independent development was occurring in China, India, and Central America. Yet, to date, the Sumerians and Egyptians get credit for being the first. Our next look will be at the ancient Greeks and how they changed the very face of mathematics.

2

The Incredible Greeks

The evolution of mathematics in the ancient world took a new twist when the Greeks arrived on the scene. Before the time of the Greeks, a knowledge of mathematics and its applications fortified the holder of such knowledge with great power. This power was not to be carelessly shared with others.

After the king or pharaoh, one of the most important individuals in an ancient kingdom was the court astronomer–mathematician. For societies surviving on farming, the critical time of the year was the planting season. If farmers planted too early, then a late frost could kill the young buds, or spring floods could wash the seeds from the fields. If they planted too late, then the crops would not mature in the fall before the first frosts ruined them. Therefore, it was critical to know the best time to plant. For this information the farmers looked to their king. In turn, the king appointed an astronomer to devise a calendar for the kingdom. To construct a reliable calendar, the astronomer turned his eyes toward the heavens.

The Earth takes slightly more than 365 days to make one complete orbit around the Sun. Thus, each day our planet moves approximately one degree in its orbit. This motion causes the stars to appear to advance each night across the sky from east to west approximately one degree in relation to the rising and setting of the sun. This is easy to see just before dawn when the various star groups—the constellations—rise above the horizon one degree more ahead of the sun than on the previous day. The seasons of the year can also be judged by watching the periodic waxing and waning of the moon, for in the eastern sky just before sunup, the moon appears as a crescent. However, if the moon appears just after sunset, it is full. This periodic cycling of the moon was so obvious to the ancients that the first calendars were generally moon calendars, relying on the fact that the average period for the moon is just under thirty days.

The astronomer–mathematicians, working in the courts of ancient kings, emperors, and pharaohs, tracked the apparent motions of constellations and the moon by recording their positions over time, and then used this information to calculate accurate calendars. Mathematics was crucial to this endeavor, for the mathematician used the recorded history of the heavenly bodies to compute the current date and then the proper dates for the coming year. Hence, mathematics and astronomy worked hand in hand to insure the smooth running of the kingdom. If the astronomers were wrong in calculating the spring equinox, disaster could befall the kingdom. Important knowledge made the astronomer–mathematician a very powerful person, so he or she would not normally share their secret calculating techniques with others. This may be one reason why so little written information has survived to tell us how mathematicians in the third and second millennium B.C. performed their calculations. All of our most ancient mathematical writings are from scribes or students, rather than from their masters. The use of the position of the stars to direct the planting of the fields led people to make a strong connection between the position of the stars and their own personal welfare—hence the birth of astrology. The advanced societies before the Greeks included a strong element of astrology mixed in with their mathematics and astronomy. Before the king would make a decision having any far-reaching impact, he usually asked the court astronomer to consult the stars for advice.

When the Greeks came along they ruined the old system of secrecy, for they wrote books about mathematics and the sciences. Anyone with the money to purchase such a book could possess the knowledge. The Greeks democratized knowledge, breaking the ancient code of secrecy. For this reason, we know a great deal about Greek mathematicians and their contributions as compared to the Babylonians, Egyptians, or Chinese.

Greek history began around 2000 B.C. when a group of nonliterate, Indo-Europeans migrated from the northern Balkans into the Eastern Mediterranean. During the next twelve centuries they established the Greek city–states. For their written language, they borrowed the Phoenician alphabet, adding vowels. By 800 B.C. they occupied between six and seven hundred cities in the eastern Mediterranean and had begun to record their epic poetry. Now they were ready to get serious about mathematics.

THALES

Thales (634–548 B.C.) lived in the city of Miletus on the west coast of Asia Minor (Figure 4). From Miletus he could travel easily to both Egypt

FIGURE 4. Thales (634–548 B.C.). Photograph from Brown Brothers, Sterling, PA.

and Babylon, learning what mathematics these societies had to offer. Neither kingdom had yet been overrun by the Persian kings but they were certainly in decline. After his travels, Thales established the first great Greek center for learning, the Ionian School.

Numerous stories have been told about Thales, but they are all secondhand, and we have no way to verify their truth. We do know that he was held in great esteem by the other Greeks, who considered him one of the seven wise men of Greece. Supposedly he made several discoveries in geometry including: a diameter bisects a circle; the angle inscribed in a semicircle is a right angle; the base angles of an isosceles triangle (a triangle with two sides equal in length) are equal; and two triangles are congruent (have the same shape and size) if they have two angles and one side that are equal. Whether he actually made these discoveries is unknown. However, he was the first human being on record to be credited with specific mathematical discoveries.

Yet Thales' theorems on geometry are not as important as his method. From the evidence of both the Babylonian clay tablets and the Egyptian papyrus scrolls, these older societies used mathematics to solve real-world problems, such as measuring areas and volumes for constructing buildings or calculating dates to maintain their calendars. When they talked of lines, they meant particular lines drawn on the ground. When they computed the volume of a cylinder, it was a real, existent cylinder. But Thales gave his lessons about abstract lines and abstract triangles. If he were to draw an isosceles triangle in the sand, he could never do it so accurately that its base angles were exactly equal. No one could draw two lines (triangle legs) with that much precision. Hence, to prove that the base angles of an isosceles triangle are equal meant he had to be talking about an abstract triangle. Therefore, the geometrical figures Thales was talking about were mental objects, objects finding a home only in his imagination. This move from the concrete objects in the physical world to ideal mental objects forever changed the face of mathematics.

When we wish to know the source of a mathematician's passion, we must look at the abstract objects of mathematics. Such objects are exact and not approximate. It is this startling exactitude contained within the mathematical relationship that so astounds us. A geometric picture or a written equation is but a poor shadow of the thing we are really contemplating. Therefore, to get into a mathematician's mind, we must learn what these wondrous objects are and how they are related to each other.

After abstracting the objects of mathematics, Thales did another most remarkable thing. He used logic to deduce his conclusions, creating the concepts of theorem and proof. He began with truths which he knew (or assumed he knew), and with the aid of logic deduced additional truths about his abstract objects. The deductive method of mathematics stands out as one of the great contributions of the ancient Greeks. Modern mathematics would be impossible without deductions and proofs. Thales established mathematics and science as independent fields of study, disassociating them from the astrology of the more ancient societies. He did this in order to establish a higher level of truth for mathematical propositions. His mathematical theorems were self-evident from their proofs and did not need to rely upon the authority of others, nor the claims of mystical insights.

PYTHAGORAS

The second great mathematician produced by the Greeks during the sixth century B.C. was Pythagoras (580–500 B.C.), who was born on the small island of Samos, just off the coast of Asia Minor (Figure 5). It is of interest to note that Pythagoras lived during the same time period as two other great thinkers: Confucius in China and Buddha in India.

Younger than Thales, he may have studied at Thales' Ionian School.

FIGURE 5. Pythagoras (580–500 B.C.). Photograph from Brown Brothers, Sterling, PA.

Pythagoras, like Thales, traveled to both Egypt and Babylon, possibly going as far as India. Afterward he established a school at the city of Croton in southern Italy. His school was also a religious cult, where a combination of religious rituals and taboos were combined with mathematical instruction. The approximately three hundred students in the school were sworn to absolute secrecy, forbidden ever to divulge the sacred teachings of their master. Because the school worked under the shroud of secrecy, some of the discoveries made by later Pythagoreans may have been incorrectly attributed to Pythagoras. The words for "philosophy," meaning a "love of wisdom," and "mathematics" may have been first used by Pythagoras. As with Thales, all we know about Pythagoras and his school comes to us secondhand.

After the death of Pythagoras, his school disbanded. Some of his disciples moved to other Greek cities and opened their own schools. Hence, the teaching of the Pythagorean society became known throughout the Greek city–states. The ideas of Pythagoras profoundly influenced later Greek thinkers, including Plato. The Pythagoreans introduced the *quadrivium* of study: geometry, arithmetic, music, and astronomy, known as the *fourfold way*. During the Middle Ages, the *trivium* of rhetoric, logic, and grammar was added to define the seven liberal arts considered essential for the well-educated individual.[1]

Pythagoras built a cult around the idea of number, his disciples' very motto being, "all is number." Numbers were not just attributes of collections of things (concrete counting) but were real existent things in themselves. He then discovered a number of magical relationships existing between numbers themselves which were not tied to any specific collection of material objects. Thus was born the mathematical discipline of number theory. Since numbers are the building blocks of the foundations of all of mathematics (studied today as set theory), the study of the relationships between these numbers is also fundamental to mathematics. Number theory deals with whole numbers, which makes it a relatively easy introduction for the new initiate. Thus, it becomes the first fascination for many young students of mathematics. Yet, many truths hidden within number theory are very deep and hard to uncover. Therefore, while many study number theory, few are lucky enough to add to its body of knowledge.

Pythagoras saw such wonderful perfection in his number relationships that he was led to believe that number must be the generator of all else. This idea may have come to him when he discovered the relation between a taut string and the tone it produced when plucked. Halve the string and the tone goes up an entire octave. The notes in between can be produced by shortening the string according to simple ratios of whole numbers. Pythagoras extended the idea of ratio and sound to other areas. As an object moves through space, he claimed, it produces a sound, and the pitch of that sound must be proportional to the object's speed. Since the planets move in perfect circles around the earth, they must also produce individual pitches proportional to their speeds. Hence was born the idea that the heavenly bodies generate a kind of music—a music of the spheres. Could this relationship between music, ratios, and proportions mean that the seemingly imperfect physical world was somehow imitating numbers or was molded by numbers?

Pythagoras went on to develop an amazing cosmology based on the idea that numbers are not only mental objects, but were the first material cause for everything in the universe. If we are to know about our universe, then we must first study numbers, which will reveal all truths. Pythagoras therefore elevated the study of numbers to the highest level. We can see the influence of this in the writings of Plato, who, in turn, influenced much of western thought.

Just what discoveries between whole numbers were made by the Pythagoreans? Some are elementary, yet signal deep mysteries. The early Pythagoreans studied numbers by manipulating small pebbles on the ground. In fact, the Greek word for pebble was *pséphoi*, whence we derive our word "calculate." Certain numbers of pebbles, they discovered, could be arranged in triangles, and were called triangular numbers. To find consecutive triangular numbers we need only add consecutive integers. Hence, we have $1 + 2 = 3$, and $1 + 2 + 3 = 6$, with both 3 and 6 being triangular. They discovered that sets of pebbles numbering 4, 9, and 16 could each be arranged into the shape of a square. Hence, squares of numbers (n^2) became the square numbers. Square numbers were also the sum of consecutive odd numbers: $1 + 3 = 4$ and $1 + 3 + 5 = 9$.

Some of this may seem unsophisticated to us now, but this kind of investigation led the Greeks into very deep and fundamental areas. They discovered that all numbers (except 1) are either primes or composites. The study of prime numbers and their distribution within the number system is one of the deepest mysteries even today. All numbers decompose (factor) into primes or are primes themselves, making primes the essential building blocks of the number system. If we are to understand the number system, we must understand primes. Simple conjectures about primes are still unanswered. For example, some primes differ by only two numbers, such as 11 and 13, or 17 and 19, and are called prime pairs. Do there exist an infinite number of prime pairs? (Twin prime conjecture). No one knows. Can every even number greater than 4 be represented as the sum of two primes? (Goldbach's conjecture). Again, no one knows.

The abstraction of numbers by Thales and Pythagoras had a strange twist. Both the Egyptians and Sumerians had expanded the number system to include fractions, yet these fractions represented parts of objects or a subpart of a unit length. When the Greeks idealized the concept of number, they did so with only the positive whole numbers. Therefore, within their idealized number system there were no fractions! Instead of fractions they talked of ratios between whole numbers. Hence, while they abstracted the objects of mathematics and introduced mathematical proof through deductive logic, they failed to expand the number system or even to incorporate fractions into their system.

This brings us to two other discoveries made by the Greeks, and probably both made specifically by the Pythagoreans. The first concerns the length of the diagonal of a unit square (a square whose sides are each equal to 1). The legend says one Pythagorean proved that the ratio between this diagonal and the side could not be written as the ratio of two whole numbers. The Pythagorean theorem can be written as:

> $A^2 + B^2 = C^2$ where A and B are the two legs of a right triangle, and C is the hypotenuse.

Drawing a diagonal in a unit square produces a right triangle where the diagonal is the hypotenuse. Therefore, we can use the Pythagorean theorem to write the length of the diagonal of a unit square as $\sqrt{2}$. By drawing a unit square, we can then draw a diagonal of length $\sqrt{2}$, a length which cannot be represented as the ratio of two whole numbers.

This was a great embarrassment to the Pythagoreans, because they had been teaching that all reality is based on whole numbers and the ratios between whole numbers. Suddenly, here was a length, $\sqrt{2}$, which violated that tenet. But theirs was a secret society, so they just swore everyone to silence—no one would find out about the proof, which they called the incommensurability of the diagonal. The legend then says that a Pythagorean, Hippasus of Metapontum, blabbed the secret to those outside the

Pythagorean society, and the truth was finally out. For his indiscretion, Hippasus was supposedly drowned by his fellow Pythagoreans in a lake. Another version of the story says that he was lost at sea in a storm brought on by the gods who were angry with him for leaking the secret.

Whatever the case, the Pythagoreans had stumbled upon a new kind of number, but a number they completely failed to see as such. For our purposes, we will call this number a surd. A surd includes the square root of a number which is not already a perfect square. For example, $\sqrt{5}$ is a surd but $\sqrt{4}$ is not because it is exactly equal to 2. We also want to include cube roots ($\sqrt[3]{n}$), and fourth roots ($\sqrt[4]{n}$), and all the nth roots of numbers.

KINDS OF NUMBERS

Surds are solutions to certain polynomials. From your high school algebra you will remember that a polynomial is an expression involving some unknown, say x, multiplied in different combinations with numbers (coefficients) and all added together. We can write the general form of a polynomial as:

$$a_n x^n + a_{n-1} x^{n-1} + a_{n-2} x^{n-2} + \ldots + a_1 x + a_0.$$

This is a rather daunting expression, so we really need an example. Consider the polynomial $3x^4 - 2x^3 + 11x^2 - x + 9$. We see that the unknown, x, is written with exponents, and that the magnitude of these exponents is in descending order, beginning with x^4 on the left to the constant, 9, on the right. Each unknown is multiplied by a number called a coefficient. The coefficient for the plain x term is just 1. The largest exponent on the x is the degree of the polynomial. Polynomials are of special interest to mathematicians. If we carefully examine any polynomial we will discover that the only operations being used are addition, subtraction, multiplication, and division. Therefore, if we restrict ourselves to using these four basic operations upon numbers (including unknowns) we always get a polynomial. Polynomials represent one kind of object that mathematicians call functions. Later we will talk more about functions.

Why all this effort to define polynomials? If we look at all polynomial equations that have integers or fractions for coefficients and find those (real) numbers which are their solutions, we will get all the integers, fractions, and surds. We say a number is a solution (or root) of a polynomial when its substitution for each x in the polynomial results in the polynomial being equal to zero. For example, the surd, $\sqrt{5}$, is the solution to the simple polynomial, $x^2 - 5 = 0$. The surd $\sqrt[3]{2}$ is the solution to the polynomial $x^3 - 2 = 0$. Notice that we used the term "real numbers" above. The real numbers are all the numbers that can be found on a number line (Figure 6). When we combine all the positive whole numbers with the negative whole numbers plus zero, we get the set of numbers called the integers. When we combine all the integers and all the fractions, we get that set of numbers called the rational numbers. All rational numbers can be written as a/b where a and b are integers (b cannot be equal to zero). Thus $3/7$, $5/1$, and $177/10223$ are examples of rational numbers.

Any number that is not rational (cannot be written in the form of a/b) is called irrational. Therefore, the surds, e.g. $\sqrt{7}$, $\sqrt{3}$, $\sqrt[3]{5}$, are also counted as irrational numbers. Do there exist other irrational numbers besides the surds? We'll come back to this question later. If we combine all the rational numbers with the surds we get the algebraic numbers. The algebraic numbers are all the real number solutions to polynomial equations that have integer or fractional coefficients. We still have more kinds of numbers to talk about, but to keep everything understandable, let's use the diagram in Figure 7 to keep things straight.

The Egyptians knew about the positive integers and positive unit fractions. The Babylonians knew about the positive integers and fractions of the form $n/60$, $n/60^2$, and $n/60^3$, etc., where n was a given integer. The Greeks knew about the positive integers, but did not consider fractions to be true numbers. They also proved that the length $\sqrt{2}$ could not be written as a fraction and was therefore a surd (irrational). However, this so frightened them, that they refused to accept $\sqrt{2}$ as a legitimate number. Instead, they separated the discipline of algebra, which dealt with numbers, from geometry, which dealt with distances (called magnitudes) as opposed to

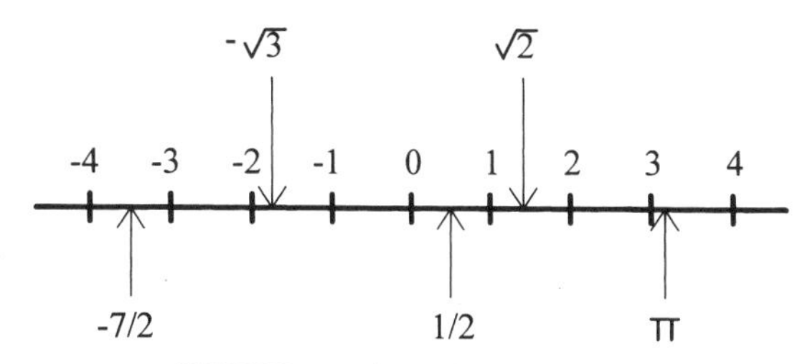

FIGURE 6. The real number line.

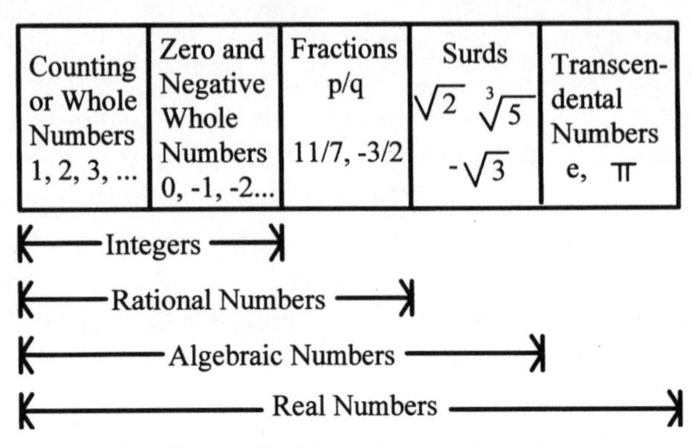

FIGURE 7. The hierarchy of real numbers.

numbers. They had to make this separation because their geometry contained lengths which could not be represented by their numbering system. Almost two thousand years would pass before geometry and algebra were put back together.

Another interesting discovery by the ancient Greeks was a special ratio between two geometric lengths. The easiest way to construct this ratio is to build a rectangle which has a width of 1 and a length of 2 (Figure 8). Now we do the same thing we did with the unit square; we draw a diagonal. The ratio that interested the Greeks was the length of the diagonal plus the width, all divided by the length. We represent this ratio algebraically as $(\sqrt{\ } + 1)/2$. This ratio can be found in the sacred symbol used by the Pythagorean school: the pentagram (Figure 9). The straight lines of the pentagram intersect each other in two places, shown as B and C

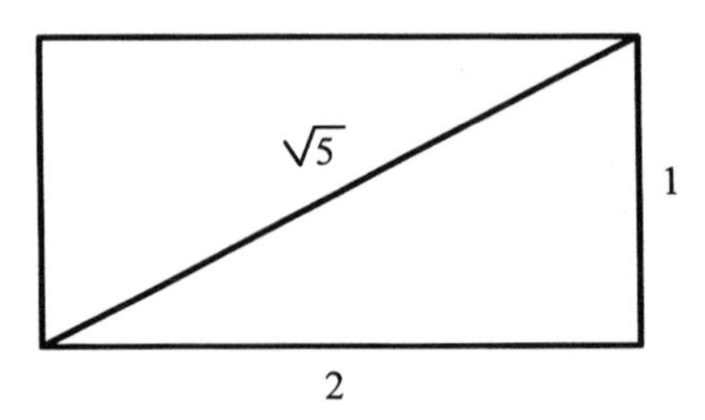

FIGURE 8. 1×2 rectangle used to build the Golden Mean.

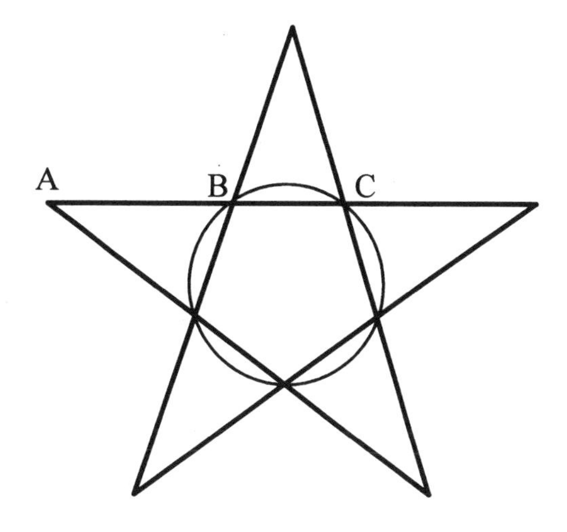

AB/BC = Golden Mean

FIGURE 9. The Pythagorean pentagram. The length AB divided by the length BC yields the Golden Mean.

in Figure 9. Now we can ask, what is the ratio between the lengths AB and BC? It turns out to be $(\sqrt{5} + 1)/2$. This ratio has many remarkable properties and was called "the division of a segment into mean and extreme ratio" or because it was so well known in ancient Greece, just "the section." The name we use today is the "golden section," "golden mean," or "golden ratio."

One extraordinary feature of the golden mean is that a line segment divided by this ratio can be folded back upon itself to replicate the ratio over and over again (Figure 10). The golden mean in decimal form is approximately 1.618 and is designated by the Greek letter phi (ϕ). Loving this ratio, the Greeks incorporated it into some of their architecture.

The golden mean has so many remarkable properties we could devote an entire book to it. Here we present only a few. First, the golden mean is one of only two numbers that generates its own square by adding the number 1 or: $\phi^2 = \phi + 1$. We can generalize this with the following identity: $\phi^n = \phi^{n-1} + \phi^{n-2}$. We will point out some really amazing characteristics of the golden mean when we look at infinite sequences and series.

Now we detour a bit more to consider an incredible coincidence. As

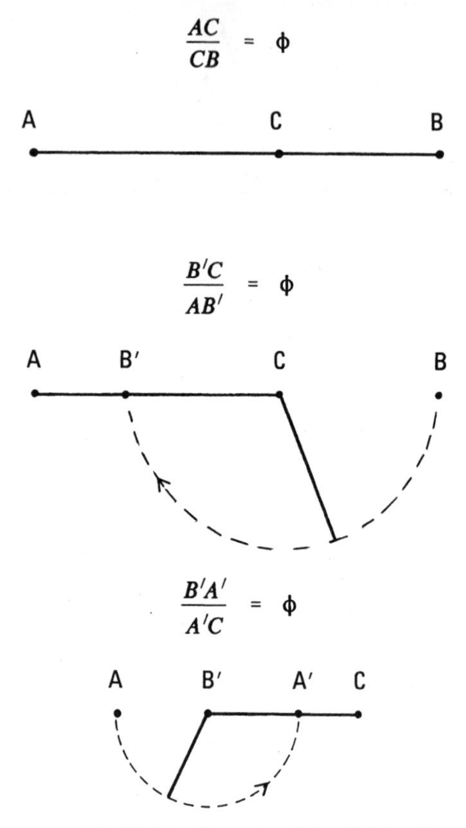

FIGURE 10. A line segment divided in the ratio of the Golden Mean can be folded back to replicate the ratio indefinitely.

far as we know, the Greeks were the first to define the golden mean. However there is an intriguing relationship in the Great Pyramid of Cheops (Khufu), built in 2560 B.C., 2000 years before the Greeks discovered ϕ. Consider the triangle that forms one of the four faces of the Great Pyramid. Draw a line down from the top of the face until it bisects the base (Figure 11). This bisecting line is 611¾ feet long. Half the length of the base is 378 feet. If we divide the bisecting line by half the length of the base we get a decimal that is remarkably close to the golden mean, in error only about 2/100 of 1 percent. This error is much smaller than our error in measuring the overall dimensions of the Great Pyramid. Is this only a coincidence or was the golden mean deliberately designed into the pyramid? We can't

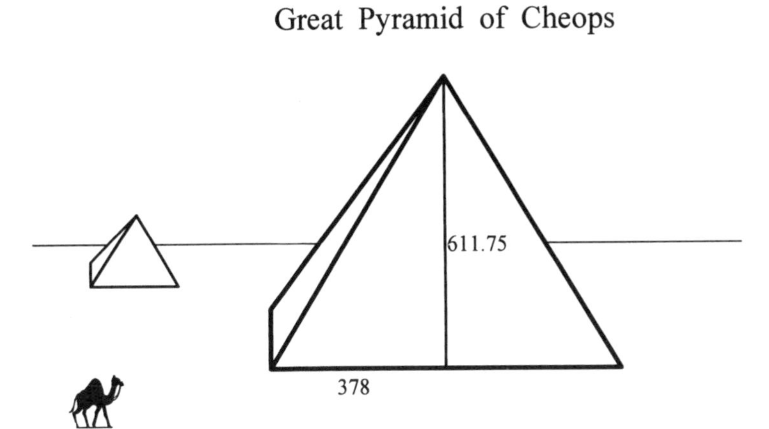

FIGURE 11. The Golden Mean and the Great Pyramid of Khufu. The bisection line of the face (611.75 feet) divided by half the base (378 feet) is a close approximation to the Golden Mean.

really say. What we need is some independent verification that the designer knew this ratio and deliberately used it. Another nice ratio is that between the height of the pyramid (481 feet) and half the length of the base (378 feet), which is approximately the square root of the golden mean or $\sqrt{\phi}$.

Now for a really glorious coincidence. The perimeter of the Great Pyramid is 3,024 feet while the height is 481 feet. Divide the perimeter by the height and we get approximately 6.28690, which is very close to 2 × pi or 6.28319, an error of approximately 6/100 of 1 percent. A coincidence? We just don't know. None of the other pyramids repeat this ratio, so we are left with no clues to the riddle.

ZENO AND THE INFINITE

The next Greek contributor we must consider is the philosopher/mathematician Zeno of Elea (fl. fifth century B.C.). Born in Elea, southern Italy, he was a student of the philosopher Parmenides, who founded the Eleatic philosophical school. Parmenides believed that reality was a single oneness, undifferentiated and unmoving, which he called the One. It was the illusion of our senses that individual objects existed and that motion acted upon them.

> Being has no coming-into-being and no destruction, for it is whole of limb, without motion, and without end. And it never Was, nor Will Be, because it Is now, a Whole all together, One, continuous; . . . Nor is Being divisible . . . Necessity holds it in the bonds of a Limit, which constrains it round about, because it is decreed by divine law that Being shall not be without boundary. For it is lacking; but if it were (*spatially infinite*), it would be lacking everything.[2]

In his attempt to support Parmenides' metaphysics of the One, Zeno produced some interesting arguments that motion was impossible. These arguments are important to us for they strike at the very core of mathematics by dealing with the infinite.

It is unclear just when people began to consider the idea of infinity. Certainly, they were thinking about it before the Greeks because the Sumerians were familiar with the concept. In the Sumerian legend of Gilgamesh, which comes to us on cuneiform tablets dating as far back as 2000 B.C., we have the wild man, Enkidu, describing hell:

> The house where the dead dwell in total darkness,
> Where they drink and eat stone,
> Where they wear feathers like birds,
> Where no light ever invades their everlasting darkness.[3]

Here we take the word "everlasting" to mean "going on without end" or infinite in time. In the Sumerian language, the word "imin" stood for seven, but also took the meaning of innumerable.[4] Perhaps the idea of infinity goes back to times before recorded history. It is not inconceivable that humans, realizing the apparent finality of death, wondered if it was possible to live forever. Many of the early gods were distinguished from humans by their immortality. Could time be infinite, or did time, like living things, die too? Certainly, by the time of the Greeks the notion of infinity was well entrenched within the human psyche. However, it was an idea that gave both philosophers and mathematicians the willies. The Greeks left us the first meaningful legacy of literature about the infinite. Most of them rejected the infinite in either the corporeal world, e.g. physical infinities, or in the world of ideas.

Aristotle is the best chronicler of the infinite for his time. He recounts Zeno's arguments against motion:

> Zeno's arguments about motion, which cause so much disquietude to those who try to solve the problems that they present, are four in number. The first asserts the non-existence of motion on the ground that that which

is in locomotion must arrive at the half-way stage before it arrives at the goal.[5]

Zeno's argument is that an arrow must go half way to the target before it can reach the target. Before it can reach the halfway mark, it must reach the quarter mark, and so forth. Hence, we are led to the conclusion that this process of division can continue forever, defining an infinite number of "halfway" marks. But, according to the Eleatic logic, a body cannot travel in a finite time over an infinite number of magnitudes. Therefore, movement is impossible. Zeno's arguments have been dealt with by Aristotle and others, but what we must stop to consider is that the idea of infinity, by the time of the ancient Greeks, was formally introduced into their philosophies and mathematics. Aristotle could not abide the idea of either a physical infinity, such as an infinite universe, nor an idealized infinity, such as an infinity of numbers. "Nor can number taken in abstraction be infinite, for number or that which has numbers is numerable."[6]

Yet, we must have the infinite in mathematics since the very foundations of higher mathematics depends upon this wonderful concept. In the work of the Greeks we see the first documented attempt by humankind to inject the infinite into mathematics, but the mystery surrounding the very notion of the infinite forced the great thinkers of that time to step back and reject the notion. This attempt to extend mathematics into the infinite is our third driving force of mathematics.

THE MYSTERIOUS LINE

Before proceeding we must pause and consider the number line. Just what is a line, anyway? From our early exposure to the concept of a line in elementary school, most of us have the idea that a line is a straight pencil mark or pen mark we draw on paper, which, we are told, is the shortest distance between two points. However, our teachers may have informed us that the mark on the paper was not what is meant by a true line, for a true line is imaginary. It is a mental object, and the mark on the paper is only a rough approximation meant to help us think about the mental construction. Just as a triangle drawn on paper is only a rough suggestion of the triangle we imagine, so with the line, the pencil mark is only a rough suggestion of the true thing. Any line we actually construct in space will be, to some degree, inexact. Our teachers would explain that our imaginary lines are composed of imaginary points—mental objects which have neither height nor width. The dots we put down on paper are not real points, but only marks to help us think about real (but imaginary) points.

And this is probably where many of us ended our discussion about points and lines. We may have learned that lines must have an infinite number of points, because we can always find a new point between two other points. This process of finding an additional point between two given points can then go on forever. For most people this idea of lines and points is adequate. After all, why are we going to worry about points and lines if we can draw their representations close enough to build airplanes, skyscrapers, and power-generating dams?

But the Greeks thought very hard about lines and points and came up with a puzzling paradox. They wanted to know just what lines and points were, for they were struggling to discover the structure of the universe. For this they had to understand how a line in space was constructed and then worry about the corresponding mental picture of such a line. The Greeks had great trouble thinking this idea through, and confusion plagued most of the thinkers who followed the Greeks down to the present century.

The Pythagoreans, representing the earliest Greek mathematicians, believed that lines extended in space were comprised of indivisible points having magnitude. Pythagorean cosmology held that in the beginning of time there existed only the One, a Monad, a kind of super granddaddy number surrounded by the Unlimited—space. In the beginning, the One separated and inhaled the space. Thus began the creation of physical numbers which exist in space and time and have magnitude. These number monads, or points, were organized according to the principle of harmony, which for the Pythagoreans was a correct number ratio. Hence, the number-atoms formed during creation must be arranged in space separated by distances which were whole number ratios of each other.

To illustrate, we can look at a two-dimensional number consisting of three monads (together representing the number 3) and forming a right triangle with equal legs. Because this atomic number is formed according to harmony, the distances between the three monads must represent unit distances (Figure 12). However, the very proof that Hippasus was supposedly drowned for demonstrates that if the length of the two legs are equal and their lengths are made up of the same whole number of unit lengths, then the diagonal must be separated by a number of units plus some fraction of a unit. We can make the units on the two legs of the triangle even smaller and more numerous, but we will never force that

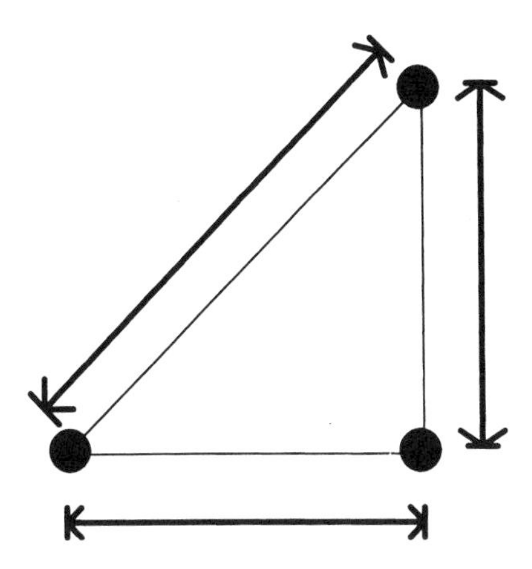

FIGURE 12. Pythagorean number monad for three. According to the Pythagorean theory of harmony, the lengths of the sides always must be in whole number ratios. This was proven impossible by the proof of the incommensurability of the diagonal of the square.

diagonal of the triangle to be the exact length of some whole number of units, i.e. some fraction of a unit will always be needed to complete the distance on the diagonal. This is why the incommensurability of the diagonal was such a blow to Pythagorean cosmology; it was a counterexample to the belief that the atomic numbers were formed strictly according to whole number ratios, i.e., a harmony between whole numbers.

If distances in space are not always some integer number of basic units of line segment, then possibly a line is made up of an infinite number of points, each of which has no length. However, this is the very idea that some of Zeno's arguments were meant to refute. If a line is nothing more than an infinity of points, each of which adds zero to the overall length of the line, then an infinity of zeros is still just zero! How does the line have any length if its parts added together are zero in length?

Maybe individual points do have some tiny magnitude, a magnitude we can call infinitesimally small, or simply an infinitesimal. If so, every line segment would be infinitely long, being the sum of an infinity of fixed length points. Well, then, maybe points have magnitude, and each individual line segment contains only a finite number of points. This is what the Pythagoreans had believed, but the incommensurability of the diagonal

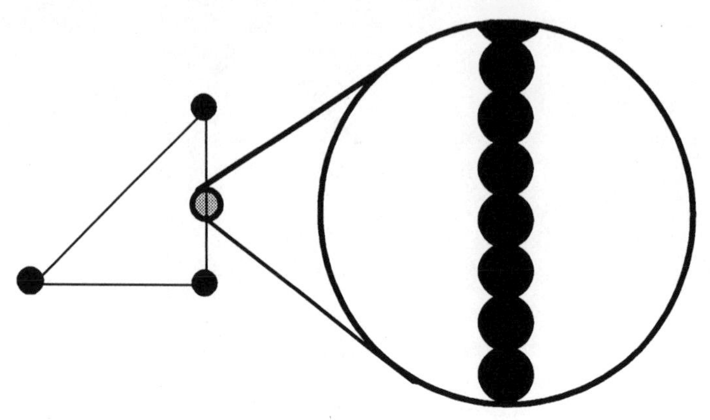

FIGURE 13. Fanciful illustration of a line constructed from points having magnitude.

destroyed that idea. That means we are back to a finite line length consisting of an infinite number of points. Maybe we can get somewhere if we just magnify a line segment and take a closer look at these mysterious points.

Figure 13 shows a blowup of the segment of line from Figure 12. We see in this fanciful illustration that the points are now large and touching each other. But of course this is nonsense, for we already have dismissed the idea that lines can be made of a finite number of points with magnitude. What we should really have is the Figure 14 which shows that the

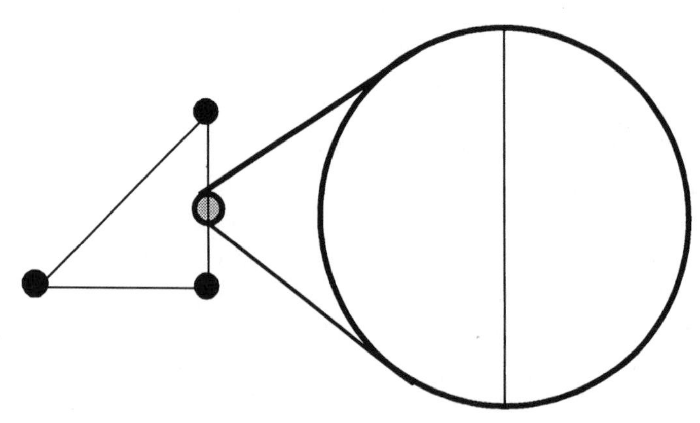

FIGURE 14. Proper magnification of a line showing no resolution into individual points.

points do not increase in size with magnification. The points shown in Figure 14 are not really points at all, but dots we write to show the approximate location of the point we wish to talk about. The point, having no dimension, must be invisible, having zero length. Now we are back to Zeno's objection that an infinite number of zero lengths add up to a zero length. What are we to do?

We are really talking about two kinds of lines: those lines that can be actually found as part of space–time and those which are purely objects of thought. The pickle we find ourselves in applies to both kinds of lines: mental and physical extension. The Greeks did not have enough numbers corresponding to points to fill up a line segment. We did not realize how many points there must be on a number line until the end of the nineteenth century when Georg Cantor proved there existed, at least as thought objects, different-sized infinities. The number line, he proposed, is composed of two kinds of numbers: algebraic and transcendental numbers. The algebraic numbers represent one example of the smallest sets of objects of infinite size—a countable infinity. Transcendental numbers are an example of the next larger infinite set, a set that is not countable. It would take these two infinite sets, each set of a different size infinity, the algebraic and transcendental, to contain enough points to "fill" the number line.

However, this does not help our intuitive sense that something is missing from a line segment if it contains nothing but dimensionless points, for an infinity of nothings is still nothing. How can we move a definite distance from one place to another if we move through an infinity of points, all of which contribute not at all to our motion, since they are dimensionless? Consider a white swan as she slowly floats through the cool waters of a mountain lake. She is in motion and passes from point to point. Between these points, there exist an infinity of dimensionless points. Zeno would have said that when the swan is at a specific position, she must be at rest, for a single one of her points cannot occupy two corresponding lake points marking her motion simultaneously. Being at rest at each of an infinity of points, she cannot move at all. Yet, she does.

Something else must be going on. Our swan must have motion at every point along her line of motion. That is, at each point she is as if still, motionless, yet she has the property of motion. We might want to call this instantaneous motion. It will become a critical idea many years after the Greeks, and an idea necessary before the complete logical development of calculus.

Before we leave this whole perplexing idea of what constitutes a number line, we must consider what a physical distance, i.e., line, in space–time is composed of. Our first guess might be a line similar to the number line of thought, which is composed of our infinite dimensionless points, yet the current state of modern physics suggests otherwise. Our best description of events at the subatomic level now relies on quantum theory. But in quantum theory events take place in discrete units. The electron, occupying a specific shell around an atom's nucleus, does not go to the next shell by moving through the space between shells. Rather, it disappears from the inner shell and instantaneously appears in the outer shell. How is this possible, unless objects at the subatomic level move, not in the smooth motion of the swan, but jerking about in infinitesimal units of equal length? Were the ancient Pythagoreans correct after all?

THE ELEMENTS OF EUCLID

Let us turn to the greatest mathematics book ever written—*The Elements* by Euclid (365–300 B.C.). But before we do, we'll want to consider the work of Eudoxus (408–355 B.C.), who lived just before Euclid's time. Eudoxus was born in Cnidus on Asia Minor and became a student of Plato. Eudoxus was also an astronomer and gave one of the first explanations of the motions of the heavenly bodies through a system of complicated rotating spheres. The ancients had recorded that the fixed stars, as they progressed across the heavens each night, appeared in slightly advanced locations. However, some "stars" seemed to move forward a little faster than the great mass of stars, and then suddenly these same stars stopped in the sky, and then actually moved backward (retrograde) against the background of stars. These strange stars were dubbed the wandering stars, or planets. How do we explain the strange behavior of these wandering stars if we assume the world view of that time of a universe that placed the earth at the center and the vault of the heavens revolving around it? Thus we see an attempt by Eudoxus to use a mathematical model consisting of spheres to explain these perplexing motions. Eudoxus' effort explained part of the retrograde motion of the planets, but not all. However, 450 years later the Greek, Ptolemy, improved on Eudoxus' method to account for a great part of the retrograde motions, thus saving the earth-centered view of the universe. Eudoxus' contribution was his attempt to use the power of mathematics to predict the behavior of the universe he dwelled in.

Some philosophers, called intuitionists, believe that mathematics is completely an invention of humankind. It is like a great logical chess game that has been devised by humans, and if we were not around, then mathematics would cease to exist. This argument has a strong appeal since it solves certain philosophical problems by just denying that the problems exist. Yet, if mathematics is just a game that humans have invented, then why does it work so well to predict events in our universe? We do not see the universe working according to the rules of chess or according to the syntax of the English language, both recognized as strictly human inventions. However, we see everywhere that the universe appears to work according to mathematical rules. It would almost seem that nature is stealing her ideas of how to act from mathematics. This ability to model reality will be considered later when we see it in the remarkable ability to predict such wondrous phenomenon as the course of the planets about the sun and the periodic change in the earth's tides.

Eudoxus is also credited with the distinction of discovering that the year is ¼ day longer than 365 days (the actual figure is closer to $\frac{11}{50}$ of a day). He is also thought to have developed a significant part of that geometry which Euclid included in *The Elements*. No mathematical books of Eudoxus have survived, even though it was reported by other ancient Greeks that he produced several. Other Greek mathematicians gave him credit for establishing a number of general theorems in both geometry and number theory.[7]

While Euclid is one of the greatest ancient mathematicians, we know almost nothing of his life. After Alexander the Great conquered Egypt, he ordered the building of a new city on the Egyptian delta, named after himself—Alexandria. When Alexander died in 322 B.C., his kingdom was divided between three of his generals, with Egypt going to Ptolemy. Ptolemy was wise enough to invite great thinkers to his new capitol of Alexandria and to establish an academy and library there. Euclid was one of the first to head this academy. Euclid, a great organizer and logician, took the current mathematical knowledge of his day and condensed it into a manuscript consisting of thirteen books. Nine of those books deal with plane and solid geometry, three deal with number theory, and one, Book 10, deals with the Greek attempt to handle incommensurable lengths. The Pythagoreans, by discovering the incommensurability of the diagonal, demonstrated that certain lengths, e.g. $\sqrt{2}$, have no corresponding numbers. Therefore, geometry (the study of lengths) had to be separated from algebra (the study of numbers). However, because such lengths as $\sqrt{2}$ occured frequently in geometry, the geometers had to develop a theory to compare different incommensurable lengths. This theory was developed by Eudoxus and written by Euclid into Book 10 of *The Elements*.

In the beginning of *The Elements*, Euclid gave a number of definitions, five postulates (truths assumed about lines and points), and five axioms (general truths). Modern logicians would dismiss any meaningful distinction between his postulates and axioms and consider all either postulates or axioms. From the definitions and postulates (axioms) Euclid logically deduced all of his theorems of geometry. This was a monumental achievement and served as a model for all of mathematics right up to the twentieth century. *The Elements* was used as a geometry text for two thousand years!

What is surprising about the geometry of Euclid, and of the mathematics of the ancients in general, is that they did not have symbolic algebra to express their propositions, yet they were able to achieve such a high level of rich mathematics. Symbolic algebra is the modern language we use (that many of us struggled so hard in high school to learn) to solve geometric and algebraic problems. It is really the special language of mathematics. Since the ancients didn't have this language, they were reduced to drawing a diagram and then stating their problem in Greek. Instead of manipulating symbols, they would stare at the diagram, draw some more pictures in the sand, and think very hard about the solution. Once they found the solution, they would try to justify it with words. We see this everywhere in the various books of *The Elements*. For example, Proposition 11 of Book 9 says:

> If as many numbers as we please beginning from an unit be in continued proportion, the less measures the greater according to some one of the numbers which have place among the proportional numbers.[8]

One can appreciate that it is not immediately obvious what Euclid is trying to get across. However, if we translate this into symbolic algebra, the meaning becomes apparent. When Euclid talks about numbers being in continued proportion, he is talking about a geometrical progression, for example: 1, 2, 4, 8, 16, 32, . . . , where each succeeding term, beginning with 1, is simply multiplied by 2. We can construct such a geometrical progression using any number. For example 1, 1.2, 1.44, 1.728, 2.0736, . . . is a geometrical progression where each number, beginning with 1, is multiplied by 1.2. These kinds of progressions can be written as $1, a^1, a^2, a^3, a^4, \ldots$

Proposition 11 states that if we take any term, say a^m, and divide it by another term occurring earlier in the sequence, say a^n, then the result will be one of the terms in the progression. We see this is true at once since $a^m / a^n = a^{m-n}$, which will be a term in the progression if $m > n$. For example, suppose we have the sequence of powers of three:

$$1, 3, 9, 27, 81, 243, 729, \ldots$$

Now if we take one term of the sequence and divide by an earlier term, the result will also be a term in the sequence. If we divide 243 by 27 the answer is 9—our third term.

It is easy for us to state the progression concept in symbolic algebra, but it was difficult for Euclid to state it in the nonmathematical language of ancient Greece. Euclid, as one of the greatest mathematicians to have ever lived, could write short, beautiful propositions, too:

> If two circles cut one another, they will not have the same centre.[9]

This is Proposition 5 in Book 3. At once we can visualize what he means: if two distinct circles intersect, then they can't have the same center. On the other hand, we have one of my favorites—Proposition 4 of Book 5, which states:

> If a first magnitude have to a second the same ratio as a third to a fourth, any equimultiples whatever of the first and third will also have the same ratio to any equimultiples whatever of the second and fourth respectively, taken in corresponding order.[10]

To understand this rather oblique statement we will rephrase it in modern terminology. We can write the following two fractions as equal to each other since they are both exactly equal to $\frac{1}{2}$: $\frac{3}{6} = \frac{7}{14}$. The four magnitudes mentioned are the four numbers where 3 is the first, 6 is the second, 7 is the third, and 14 is the fourth magnitude. Euclid's statement says that if the two fractions are really equal, then we can multiply the two numerators (3 and 7) by the same constant and at the same time multiply the two denominators (6 and 14) by another constant, and the resulting two fractions will have the same order of magnitude between their numerators and denominators. Hence, if we multiply the numerators by 5 (a constant) and the denominators by 2 (another constant) we get the two fractions:

$$\frac{5 \cdot 3}{2 \cdot 6} = \frac{5 \cdot 7}{2 \cdot 14} \rightarrow \frac{15}{12} = \frac{35}{28}$$

Now for the original fractions, $\frac{3}{6}$ and $\frac{7}{14}$, each numerator is smaller than its denominator. In the new fractions, the numerators are both larger than their corresponding denominators. Therefore, the order was reversed for both fractions by the multiplication. The original statement says that in order for the original two fractions to be equal, then every such multiplication by constants will preserve this order relationship.

We have taken this little excursion into Euclid's *The Elements* to demonstrate several points. First, we recognize the elegant work accomplished by the Greeks under such an extreme handicap. Second, it illustrates the need for an expedient notation within mathematics, which is our third driving force in mathematics—the creation of symbolic algebra.

THE *CONICS* OF APOLLONIUS

Apollonius (262–190 B.C.) was born in Pergain, located in a part of Asia Minor which is now modern Turkey, but spent much of his adult life at Alexandria where he may have both studied and taught at the Alexandrian Academy. In the ancient world he was even more revered for his geometry than Euclid and was known as "The Great Geometer." The two greatest mathematical works surviving from the Greek period are Euclid's *The Elements* and Apollonius' *Conics*. Apollonius also contributed to astronomy by introducing the idea of epicyclic motions to explain planetary motion as opposed to Eudoxus' use of concentric spheres. These systems were elaborated on by a later Greek mathematician, Claudius Ptolemy (A.D. 85–165), who promoted the earth-centered theories that dominated astronomy until Copernicus finally replaced them with a heliocentric theory in 1543.

Apollonius uses two cones stuck together at their points to generate three remarkable curves, the parabola, ellipse, and hyperbola. Before Apollonius, Greek mathematicians had discovered these three curves by slicing a plane through various cones. What Apollonius did was to show that all three could be generated using his system of two cones stuck together, and then he went on to derive many exquisite theorems about these curves. He also gave them the names we use for them today. In Figure 15 we have taken a line at right angles to the ground. We call this line the axis of our cones. We then take a second line, the generating line, and rotate it at a constant angle around the axis. Notice that in Figure 15 this rotation generates two cones joined at their points. The results are the cones in Figure 16 which we call right cones because their axis is perpendicular to the ground.

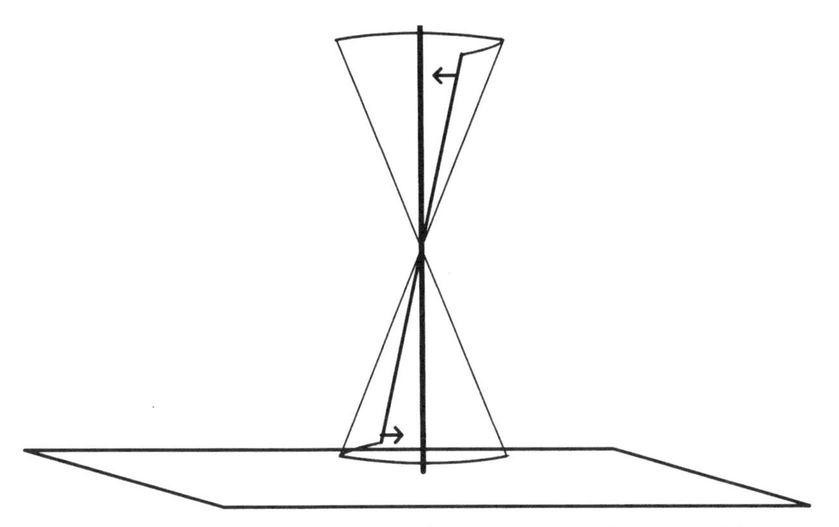

FIGURE 15. Two right cones are generated by rotating a tilted line around the axis.

FIGURE 16. Two right cones used to generate the conic sections: the ellipse, parabola, and hyperbola.

We can now generate the three curves we are after by taking a plane and slicing through these two cones at different angles. Apollonius actually showed that the curves could be generated by slicing through various cones including those that are not right cones, called oblique cones, and that we still get the three conic curves.

First we generate a circle by slicing through just one of the two cones parallel to the ground as in Figure 17. The circle is really a special case of an ellipse. Next we generate a standard ellipse by tilting the slicing plane as in Figure 18. We continue to tilt the slicing plane to generate a parabola by tilting the plane until it is parallel to the side of one cone, but cutting through the second (Figure 19). Now if we continue tilting the plane, it will intersect both cones and generate a curve consisting of two separate branches called a hyperbola (Figure 20).

Now that we have generated these curves, we should take a moment to appreciate them as presented in Figure 21. They really are magnificent curves, lovely to look at. In all probability, the circle was the first geometric curve to be constructed and admired by humans. The circle is truly remarkable because the degree of curvature is always the same. When you rotate a circle about its center you get the exact same circle back. This, of course, is one of the charming characteristics of a circle—it is so perfectly

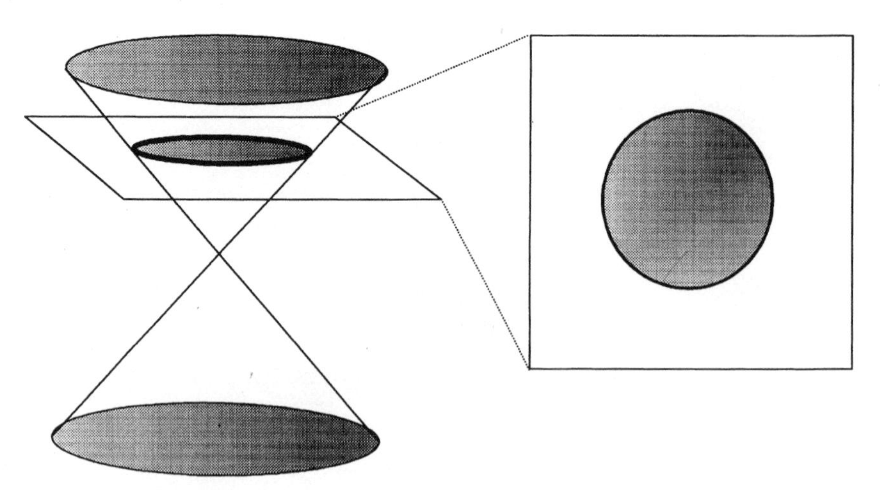

FIGURE 17. Generating a circle from two right cones. The intersecting plane cuts the cone perpendicular to its axis.

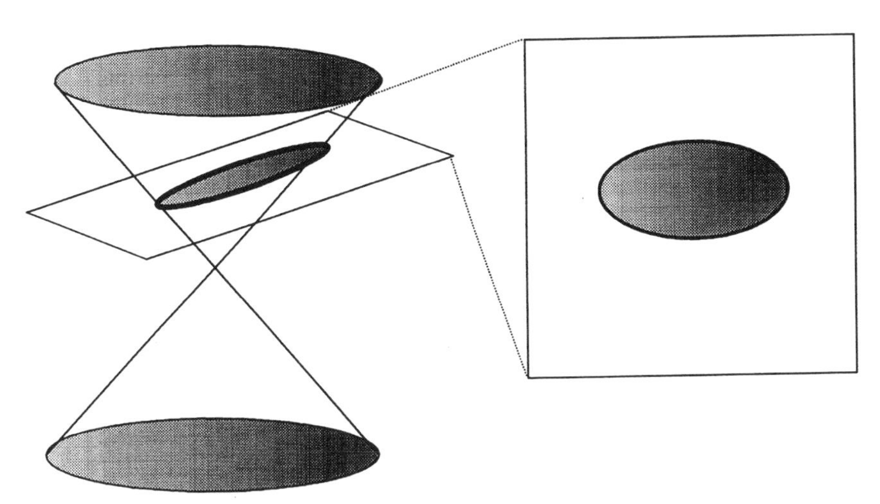

FIGURE 18. Generating an ellipse from two right cones. The intersecting plane is tilted at an angle smaller than that made by the generating line.

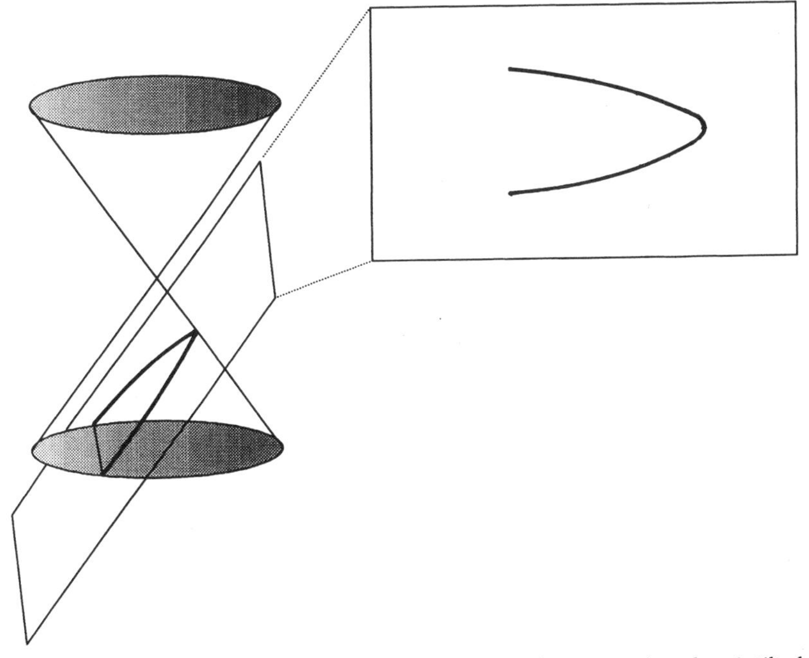

FIGURE 19. Generating a parabola from two right cones. The intersecting plane is tilted at the same angle as the generating line.

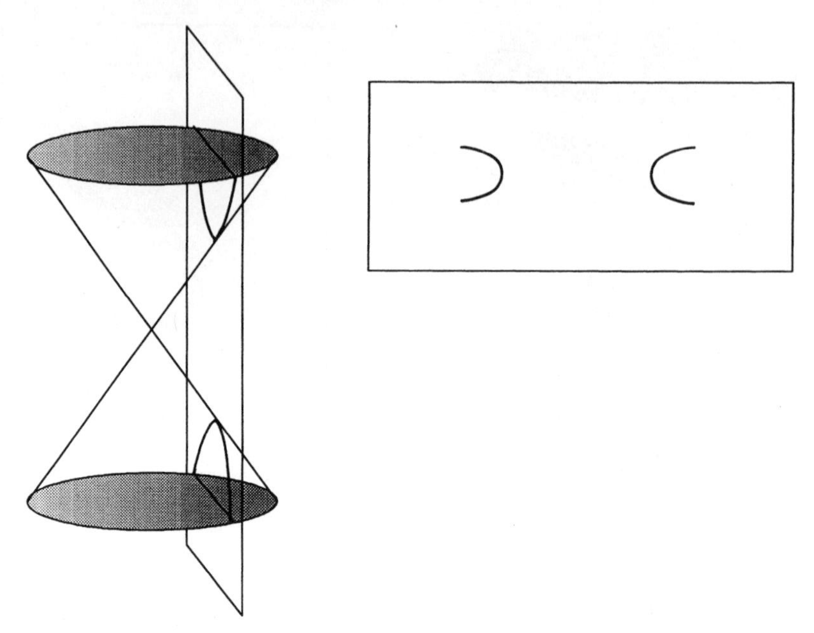

FIGURE 20. Generating a hyperbola from two right cones. The intersecting plane is tilted at an angle greater than the generating line.

round everywhere. But now, thanks to the ancient Greeks, we have three additional curves to ponder. Notice that the curvature of each conic is not constant, but is always changing. Just looking at them and enjoying them is enough, but it turns out that, unbeknownst to the Greeks, these three curves are incredibly useful. For the Greeks, a mathematical idea did not have to be useful to be interesting, since they contemplated mathematics for the sheer joy of it. When we add usefulness to a mathematical idea we get a practical utility which interests everybody.

First we should notice that the parabola is formed at only one angle by the slicing plane—when the slicing plane is parallel to the sides of the cones. Hence, we can think of the parabola as a kind of boundary between all the various ellipses and hyperbolas. Now suppose we are standing on the surface of the earth and shoot off a cannon. Soon the cannonball falls back to earth. The path the cannonball follows is a parabola (Figure 22). Any object thrown or shot into the sky (at an acute angle to the ground) that returns to earth follows the path of a parabola, except for any deviations due to air resistance.

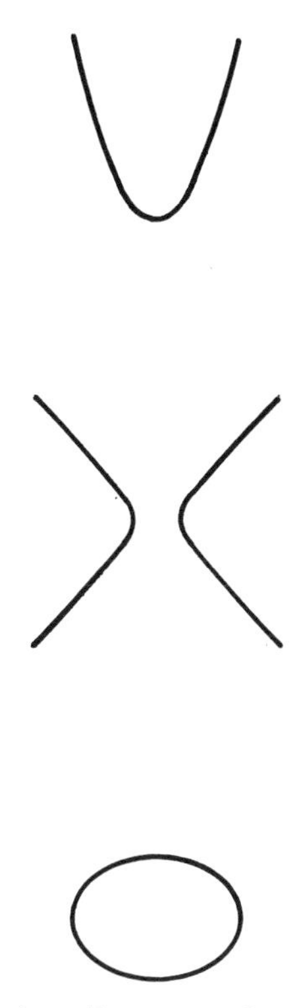

FIGURE 21. The three conic sections. From top to bottom, the parabola, hyperbola, and ellipse.

When we put a satellite into orbit around the earth, the orbit of the satellite is an ellipse. It might be a circle, but then a circle is just a special kind of ellipse. Next we take a more powerful rocket and send a space probe to fly past Jupiter. The path of the space probe is a hyperbola. Absolutely remarkable—the three paths followed by the cannonball, the satellite, and the space probe are the conic curves: the parabola, ellipse, and hyperbola. We now know that the planets move about the sun in elliptic orbits. Comets that return time and again to speed around the sun are following ellipses. If a comet passes the sun only once and never returns, then its path through the solar system is a hyperbola. These wondrous curves, first discovered by the ancient Greeks and studied by Apollonius, are the curves that bodies follow when moving within the universe. How very much we get from so little!

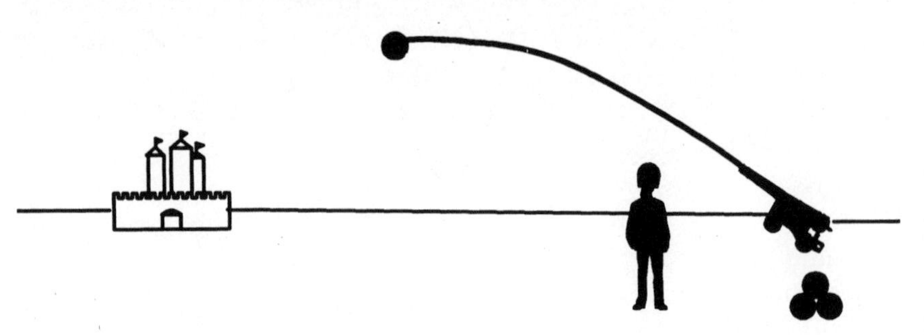

FIGURE 22. The path of a cannonball is a parabola.

Parabolas have an extra special place in our modern world, one which was never dreamed of by Apollonius. The parabola concentrates incoming light at a single point called the focus. This characteristic is used to design reflecting telescopes, such as the Hubble Telescope. Using the same law of optics, a light source placed at the focus reflects off a parabola into a beam of light. This is used in flashlights and auto headlights. Modern telecommunications uses parabolic dishes to communicate with satellites. Because these curves are so wonderful we will return again to them in later chapters.

THE GREEK HERITAGE

Before we leave the Greeks, we should mention several more who made notable contributions. Eratosthenes (276?–196? B.C.) measured the size of the earth and was off by only 15 percent. He did this by noticing that, during the summer solstice, the sun shone directly down a well in Syene, Egypt, which meant the sun was directly overhead. A year later he was in Alexandria, approximately 500 miles north of Syene, during the summer solstice. There he noticed that the sun was not directly overhead at high noon, but cast a shadow. He used this information to compute the angle of inclination, from which he computed the approximate size of the earth. If Christopher Columbus had used Eratosthenes' estimate for the size of the earth, he would have realized that China could not be a mere three thousand miles from Europe, and probably would not have sailed on his historic journey.

Hipparchus (190–120 B.C.) was born in Nicaea, a town in Asia Minor, now in modern Turkey. He worked primarily in astronomy, charting approximately a thousand different stars and discovering the precession of the equinoxes. He is especially remembered for devising a table of trigonometric cords which influenced the development of trigonometry.

Aristarchus of Samos (310?–250? B.C.) was the first to claim that the Earth revolves around the sun. Had his ideas prevailed, astronomy would have jumped ahead by 1,800 years. He devised a method to measure the distance to both Sun and moon from Earth, but came up with the wrong answers because of inadequate instruments.

We must mention Archimedes of Syracuse (287–212 B.C.). He is frequently listed as one of the three greatest mathematicians to have ever lived and is thought by many to be the greatest ancient mathematician.[11] He not only contributed to mathematics, but like many of the distinguished mathematicians, also made remarkable discoveries in the sciences. He invented a water pump based on a screw which is named after him and still used today. He was especially adapt in geometry and is reported to have such a fascination for geometrical problems and proofs that, while engaged in other activities, he would become entranced, stop what he was doing, and begin tracing geometric figures in the dirt or even upon his own body. He truly loved geometry for its own sake. One legend is that he frequently became so obsessed with work he would forget to eat or bathe, requiring his friends to forcefully remove him to the public baths.[12]

His proofs include the demonstration that the volume of a sphere circumscribed inside a cylinder is two-thirds that of the volume of the cylinder. He estimated the value of π to a high degree of accuracy by both circumscribing and inscribing a circle with polygons, using polygons with as many as ninety-six sides. This procedure became known as the method of exhaustion and was a precursor of the modern concept of a limit. Hence, he anticipated the methods used by such giants as Isaac Newton (1642–1727) and Gottfried Leibniz (1646–1716) in the invention of calculus.

During the Second Punic War, at the Roman conquest of Syracuse, Archimedes helped with the defense of his native city by constructing wondrous machines to repel the attacking Romans. He is credited with inventing a catapult used to hurl giant logs at the attackers, and with the compound pulley, used to lift and shake attacking ships in the harbor. It is even claimed that he constructed mirrors to reflect the sun onto the Roman ships, causing them to burst into flame. Reportedly, the Roman soldiers became so nervous anticipating what Archimedes would unleash on them, that the sight of a mere rope dangling from the city's wall caused them to run away, screaming "Archimedes' Machine! Archimedes' Machine!"

Finally, through treachery, the Romans overran the city. The Roman general wanted to meet this famous Greek and sent his soldiers to find Archimedes. The legend says that a soldier found Archimedes drawing figures in the sand. When told to stand and come, Archimedes brushed the soldier off and returned to his beloved drawings. The enraged soldier drew his sword and killed Archimedes.

3

Mathematical Proofs

One of the great distinguishing features of mathematics is the concept of a proof. Once again, the Greeks were the first to introduce this idea. A proof is an attempt to establish the truth of a statement, and in the case of a mathematical proof it establishes the truth of a mathematical statement. Hence, a proof is a demonstration of a statement as true. The idea of a "demonstration" is critical here.

Ancient, pre-Greek wisdom was secret wisdom. Only by joining a specific trade or cult did you receive the sacred instruction associated with that cult. Even the Pythagorean society in ancient Greece, as we saw, was a cult sworn to keeping its teachings secret. Secrecy was frequently a characteristic of alchemy during the Dark Ages. It was not until you were admitted to the alchemist's inner circle that you received instruction. And you didn't question that instruction, for it was based on ancient learning that had to be true. Occult science today still relies on vague references to "ancient knowledge." Once an ancient master discovered a truth, he simply passed it on. His disciples did not question it, for the truth came from the master. While ancient magicians conducted demonstrations of their powers, the audience was never invited to understand how the illusions were carried out.

The introduction of the idea of proof, or a demonstration of truth, by the Greeks added a new dimension first to mathematics and ultimately to science. It was no longer necessary to accept a mathematical truth because the master said it was so. It was possible to demonstrate the truth, independent of any individual, master or not. Hence, what became important was not secrecy, but public exposure. The more who saw the demonstration, the greater acceptance for the truth involved.

Proofs therefore helped to democratize knowledge because the validity of a statement rests on its demonstration and not on the say-so of someone in authority. This idea so impressed the Greeks that they believed the highest forms of knowledge were mathematics and logic, and no one could claim to be educated without having a thorough understanding of these two disciplines. Unfortunately, in today's modern climate we have slipped away from such high standards.

The first person to be given credit for a mathematical proof has already been mentioned—Thales of Miletus (634–548 B.C.), who is said to have proved five theorems in geometry. Whether Thales was the very first to use logic to demonstrate the truth of a problem in geometry is questionable. First, he traveled to both Egypt and Babylon in his youth, where he undoubtedly absorbed much from cultures which had mathematical traditions over two thousand years old. It is not a stretch of our imaginations to believe that he saw demonstrations of specific computational proofs, and that these demonstrations gave him the initial idea to generalize those demonstrations. Furthermore, we must remember that both the Egyptians and Babylonians developed procedures for carrying out specific calculations for measuring areas, or the distribution of goods, etc. In the writings that have survived, we see such methods demonstrated. The Egyptians also used the method of substitution to verify a result in a calculation. Once a calculation for a problem was completed, the mathematician substituted the answer back into the original problem to see if it made sense. This, again, is clearly a demonstration or a proof that the original calculation was correct. Therefore, the idea of demonstration was already established before the time of Thales. That such demonstrations did not meet the technical definition of a proof in modern mathematics should not be taken to mean that the ancient civilizations before the Greeks had no idea of what proofs were.[1]

Many people shy away from proofs, claiming they are too tedious and difficult to understand. "Why bother with proofs?" they say, "Just give me the facts, and I'll leave the proofs to others." However, it is the very fact that mathematics requires proofs that makes it the powerful and beautiful activity it is. We can even make a stronger statement regarding proofs. It is the concept of a proof that is at the very core of mathematics. To fully appreciate mathematics we must come to understand why this is so. Rather than beginning with technical definitions and demonstrations of mathematical proofs, let's begin with a discussion of how proofs are related to our lives and to mathematics in general.

The word "proof" and the phrase "to prove" are common in our everyday lexicon. When we use such words we have some kind of demonstration in mind. These demonstrations may be a physical act or a logical argument. Webster's dictionary defines proof as:

1. any effort, process, or operation that attempts to establish truth or fact; a test; a trial; as, to make *proof* of the truth of a statement.
2. something serving as evidence; that which proves or establishes; a convincing token or argument; a means of conviction.[2]

For example, a friend might claim that he can throw a rock across a river. By responding, "prove it," you are asking for a physical demonstration. He picks up a rock and deftly throws it to the opposite shore. You are satisfied by his act—his demonstration. Another example of a proof may be the claim that a certain politician is taking bribes. To prove this allegation, the accuser produces documents showing that the politician received goods in exchange for his political action.

Another kind of proof is logical in nature, which is the kind we are interested in. A silly joke, passed around for years, goes like this: I can prove you are not here.

1. You are not in St. Louis, nor New York, nor Denver.
2. If you are not in one of those places, you must be someplace else.
3. If you are someplace else, you can't be here!

The logic, of course, is flawed, but the overall format is sound. For a logical proof, we produce a chain of statements leading to the conclusion we are after. This method is the very nature of a mathematical proof. One condition for a mathematical proof is that the objects talked about are, in general, not people, places, etc., but the objects of mathematics, which are frequently, but not always, numbers or geometrical objects such as points and lines. Hence, a mathematical proof is a chain of statements about mathematical objects leading to a final statement that establishes a claim. We call the last statement (the claim) the conclusion, and all the preceding statements are premises. To avoid the confusion of ordinary language we generally translate the statements of the proof into symbolic algebra.

In addition to the premises and conclusions, we also have the rules of logic, which allow us to move from one premise to the next and finally to the conclusion. Originally, these rules were just commonsense notions. The ancient Greeks saw the need to look more closely at logical rules, which became formalized in Aristotle's system of logic. Aristotelian logic was the accepted bible of logic in the western world until the nineteenth century, when modern logicians began adding to Aristotle's effort. Therefore, we now have three concepts to keep in mind in connection with mathematical proofs: the premises; the conclusion; and the rules of logic used to move from premise to premise, ending with the conclusion.

Aristotle was the great synthesizer of Greek ideas regarding proofs. He recognized that the form of the premises could be used to determine whether a logical argument was valid. When we speak of statements we say they are either true or false, but when we talk of proofs, they are not true or false, but valid or invalid. One of the forms Aristotle recognized as producing a valid argument was the syllogism. Probably the most famous syllogism ever written for demonstration purposes is:

Socrates is a man
All men are mortal

Socrates is mortal

In our syllogism we have separated the two premises from the conclusion with a line. Being a valid argument form, this syllogism guarantees that if the two premises are both true (Socrates is a man and All men are mortal) then the conclusion (Socrates is mortal) must be true. A little thought convinces us that this is the case. Suppose the premises were true but the conclusion was false—that is, Socrates is immortal. This leads immediately to a contradiction since all men are mortal and Socrates is a man—hence being a man, he must be mortal.

Now let's take out the words "man," and "mortal." And replace them with the variables A and B.

Socrates is an A
All A are B

Socrates is B

In this form we see that the original argument did not rely on the attributes of "man" or "mortal." It doesn't matter what we put in for A and B, so long as the resulting premises are true, then the conclusion is true. We can even get rid of "Socrates," replacing him with the letter C.

C is an A
All A are B

C is B

Again, no matter what A, B, and C are, if, after substitution, the two premises are true, then the conclusion is guaranteed to be true. For example we will substitute the race horse, Seattle Slew, for Socrates, the noun horse for man, and the noun mammals for mortal.

Seattle Slew is a horse
All horses are mammals

Seattle Slew is a mammal

The above argument is valid.

The lack of valid proofs sometimes characterizes the mathematics of amateurs, who do not realize the importance that proofs play in the discipline. Frequently, an amateur will stumble onto a mathematical theorem, which is nothing more than a general (provable) mathematical law.[3] The amateur convinces himself that the law is true not with a proof, but by trying a number of individual cases and seeing that in each case the law holds. For example, an amateur may stumble upon the law that consecutive odd integers sum to square numbers, which we mentioned before. He will try several individual cases:

$$1 + 3 = 4 = 2^2$$
$$1 + 3 + 5 = 9 = 3^2$$
$$1 + 3 + 5 + 7 = 16 = 4^2$$

In each of these cases he has produced a number that is the next perfect square, e.g. 2^2, 3^2, and 4^2. Now he is convinced the law is true for all consecutive sums of odd integers, e.g., they always add to perfect squares. But this is clearly inadequate, because it could turn out that on his next attempt, the law would not hold.

What the professional mathematician strives for is a proof which will include the sums of any number of consecutive odd integers. A proof will demonstrate that all finite series of consecutive odd integers (beginning with one) will sum to a perfect square no matter how long the series is. The proof might follow these steps.

Clearly the law holds for the first three odd numbers. If we can prove it holds true for a series of indefinite size, then we know it will always hold true. Therefore, we first assume that the law holds for a series of n odd numbers (which is clearly true since it holds for series where $n = 1$, 2, and 3). Hence:

$$1 + 3 + \ldots (2n - 1) = n^2$$

Notice that we have found an expression for the last term, namely $(2n - 1)$, which is nothing more than the nth odd number. Now we are going to add the next odd number in the sequence to both sides, which is the odd number $[2(n+1) - 1]$. When we add the same thing to both sides of an equality, we do not violate the equality. Adding to both sides we get:

$$1 + 3 + \ldots (2n - 1) + [2(n+1) - 1] = n^2 + [2(n+1) - 1]$$

On the left we have simply added the next odd number. On the right we multiply out to eliminate the parenthesis and collect terms.

$$1 + 3 + \ldots (2n - 1) + [2(n+1) - 1] = n^2 + 2n + 1$$

The right side of the equation can be factored into a perfect square because $(n + 1)^2 = n^2 + 2n + 1$.

$$1 + 3 + \ldots (2n - 1) + [2(n+1) - 1] = (n + 1)^2$$

Hence, we have demonstrated that if we begin with a consecutive series of odd numbers and add the next odd number, then we will get the next perfect square. This demonstration is our proof that the law holds for all series of consecutive odd numbers beginning with 1.

The proof we have given is purely algebraic. A much simpler geometric proof exists, and it is this proof which convinced the Greeks. Suppose we begin with a single dot which represents the number 1, as in Figure 23. Now we can make a perfect square by adding three more dots onto the top and right sides. If we want the next square, or 9, we just add five dots to the top and right of the number 4 we have constructed in Figure 23, yielding Figure 24. We easily see that adding seven more dots (the next odd number), top and right, creates a 4×4 square representing the next perfect square number, 16. This proof convinces us that adding consecutive odd numbers of dots to the top and right of the figure will continue to generate the next perfect square.

The power of our proof is that we have demonstrated it for an infinite number of individual cases, and there clearly exist an infinite number of series of consecutive odd integers. Proofs generally have this universal flavor. They not only prove the law for a finite number of individual cases, but also for an infinite number of cases. Because of the generality of this particular proof, we never have to demonstrate a particular case, but know the law works for all cases. Hence, if someone says: what is the sum of the first 20 odd numbers, we can reply at once, 20^2 or 400. What is the sum of the first 100 odd numbers? It is 100^2 or 10,000. What power from a simple proof!

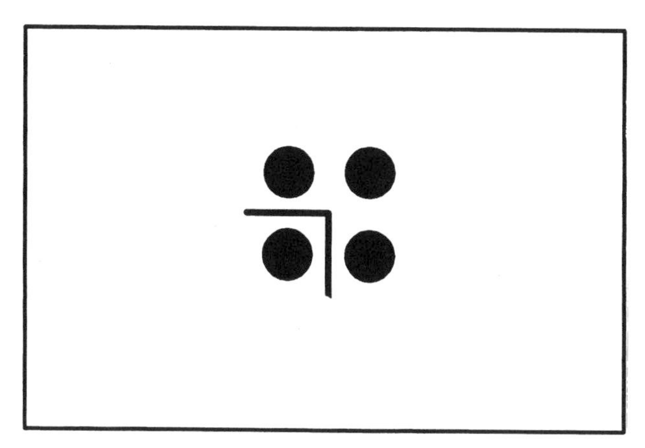

FIGURE 23. Illustrating that the sum of odd numbers yields square numbers. Adding three dots to one dot yields the square number four.

An example of a mathematical rule working for the first few cases, and then failing is the famous problem of Fermat's Numbers. Pierre de Fermat (1601–1665) was the most famous amateur mathematician to have ever lived, and we will hear more about him later. Fermat guessed that all numbers of the form:

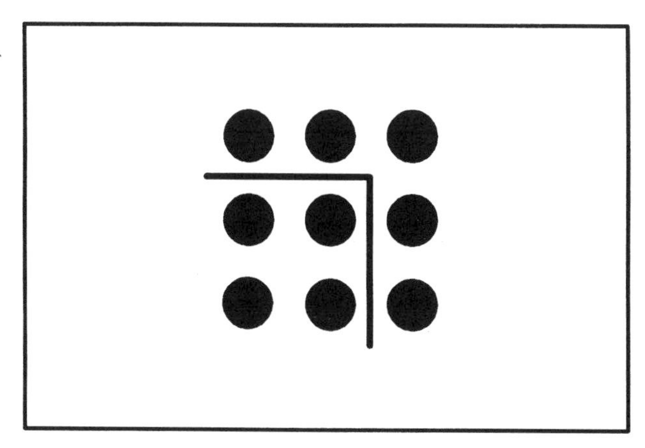

FIGURE 24. Illustrating that the sum of odd numbers yields square numbers. Adding five dots to four dots yields the square number nine.

$$F_n = 2^{2^n} + 1$$

where n is any integer zero or larger, were prime numbers. This works for the first five Fermat numbers, F_0 through F_4 since:

$$2^{2^0} + 1 = 3, \quad 2^{2^1} + 1 = 5, \quad 2^{2^2} + 1 = 17,$$
$$2^{2^3} + 1 = 257, \quad 2^{2^4} + 1 = 65537$$

It turns out that F_5 is a huge number with ten digits, and Fermat could not determine whether it was prime. However, a century later, Leonhard Euler (1707–1783) demonstrated that F_5 was composite, or $F_5 = 641 \cdot 6700417$. In fact, the very next Fermat Number after F_5 fails to be prime for $F_6 = 274177 \cdot 67280421310721$. Were F_5 and F_6 simple exceptions to a general rule? No, for all the Fermat Numbers up to and including F_{21} are known to be composite numbers.[4] F_4 is the largest Fermat Number known to be prime. Not only was Fermat's conjecture wrong about all Fermat Numbers being prime, it was outstandingly wrong. This shows how dangerous it is to generalize a mathematical law after testing only a few cases.

A mathematical proof is one of the first areas of human endeavor where we are able to deal directly with the infinite. By constructing a mathematical proof, we frequently demonstrate a mathematical law for an infinite number of cases. For such a limited effort of thought, we produce a tool of infinite domain.

Proofs come in many flavors, but two types are favorites of mathematicians: proof by construction and proof by contradiction. Proof by construction is to simply construct the mathematical object or relationship we claim exists. For our example of a constructed proof we will consider the quadrature of the lune by Hippocrates. This will not only provide an excellent example, but will demonstrate an important historical juncture.

THE MARVELOUS LUNE

The Greeks not only had difficulty understanding the basic properties of lines, but they were also struggling to measure the space confined by lines. As demonstrated by the work of Euclid, they made great strides learning about shapes constructed from straight lines, such as triangles, squares, cubes, and pyramids. Such shapes in two dimensions are polygons, and in three dimensions are polyhedron. However, they were much less successful when it came to areas bounded by curves. The circle was the first area bounded by a curve to be studied. Since geometry taught them to compare various polygons and polyhedrons to each other, certainly they could compare a circle, that most perfect yet curved figure, to polygons. Thus was born the famous problem called "squaring the circle." Given a circle of radius r, how do we construct a square with the same area?

This would seem a simple problem. The area of a circle is given by the formula $A = \pi r^2$ where π is the ratio of a circle's circumference to its diameter, and r is the circle's radius. Now a square with the same size will have one side with a length that is the square root of this or:

$$\text{length of a side} = \sqrt{\pi r^2} = r\sqrt{\pi}$$

Therefore, all we have to do is construct, using only a straight-edge and compass, a length equal to the circle's radius multiplied by the square root of π. But, we can't do it. The value π is not the solution of any polynomial with rational coefficients, and hence is not the result of any finite number of additions, subtractions, or multiplications of rational numbers. Since it is not the solution to any polynomial with rational coefficients, then it is not an algebraic number. This was not realized until as late as 1882 when π was proven to be one of those strange transcendental numbers—numbers which we need in conjunction with the algebraic numbers to fill up the number line.

So why would the Greeks believe they could square the circle in the first place? Was there any evidence that an area bounded by curves could be equated to the area of some polygon? This is where the remarkable mathematician Hippocrates of Chios enters the picture. This Hippocrates is not the famous physician, who was Hippocrates of Cos. Our Hippocrates (470–410 B.C.) was born on what is now the island of Khois, Greece. It is reported that he was a merchant who left his native island in 430 B.C. to travel to Athens. Somehow, he was robbed or swindled out of his fortune, arriving in Athens penniless. Thus, for our good fortune, he abandoned the trades to devote his life to a study of mathematics.[5] He composed a book, *Elements of Geometry*, but unfortunately no mathematical work from fifth century Greece, including Hippocrates' book, has survived. His lost book preceded *The Elements* of Euclid by at least a century.

Hippocrates was interested in the problem of squaring the circle, and his success with another similar problem may have led him to believe it was possible. The crescent shape which is trapped between two intersecting circles (Figure 25) is called a lune. Hippocrates proved that this area, an

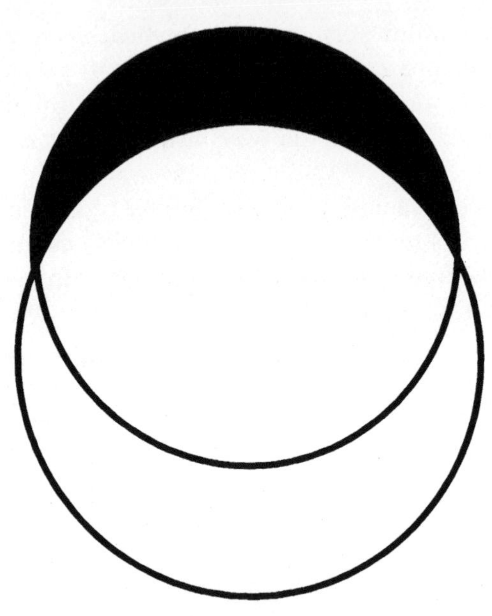

FIGURE 25. A lune is that area trapped between two intersecting circles.

area completely bounded by two curved lines, was, under certain conditions, exactly equal to the area of a specific polygon—a triangle.

Since no manuscript of Hippocrates survives we must rely on his work on lunes as reported through Eudemus in his *History of Geometry*, as recorded by Simplicius in his commentary on Aristotle's *Physics*.[6] Such circuitous reporting of the achievements of the earliest Greek mathematicians is unfortunate, but necessary. As a foundation to his proof, Hippocrates first noted that the ratio of the areas of two circles is equal to the ratio of the square of their diameters or:

$$\frac{Area\ 1}{Area\ 2} = \frac{(diameter\ 1)^2}{(diameter\ 2)^2}$$

Since the areas of circles are in ratio to the square of their diameters, he noted that the segments which are similar to each other also have areas whose ratios are equal to the ratio of the squares of their respective diameters. This is somewhat more difficult to see. A segment is that area of a circle that we trap between a straight line and part of the circle (Figure 26).

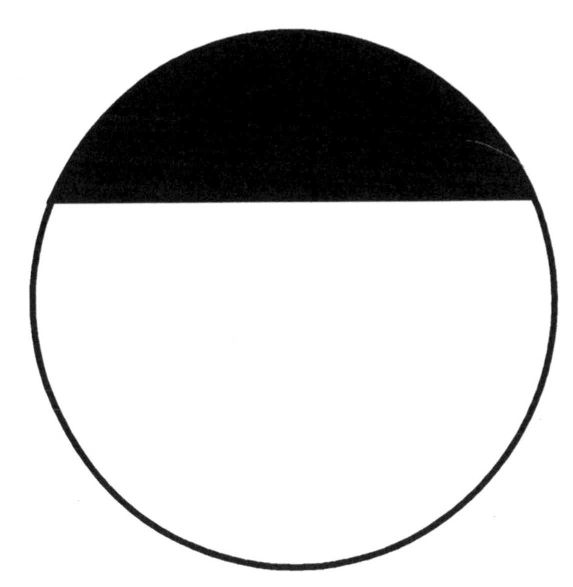

FIGURE 26. A segment of a circle is that area trapped between a straight line (a chord) and a circle.

Two segments are similar if they are formed from equal angles (Figure 27). Since the two circles in Figure 27 have areas that have a ratio equal to the ratio of the square of their diameters, then the corresponding areas of their segments will also form such a ratio. We can go one step further. The areas of two similar segments will be in a ratio that is equal to the ratio of the square of the length of the straight lines that form their boundaries. Thus we have:

$$\frac{\textit{Segment Area } 1}{\textit{Segment Area } 2} = \frac{(\textit{Segment Length } 1)^2}{(\textit{Segment Length } 2)^2}$$

With this accomplished, Hippocrates next constructed the following lune. Take a semicircle and inscribe a isosceles triangle as in Figure 28. It was known to the Greeks that the angle opposite the semicircle's diameter is a right angle, thus making the triangle a right triangle. This means that the sum of the squares of the lengths of the two legs will equal the square of the length of the hypotenuse, which in our case is the diameter of our semicircle. The two legs of the right triangle now form two segments.

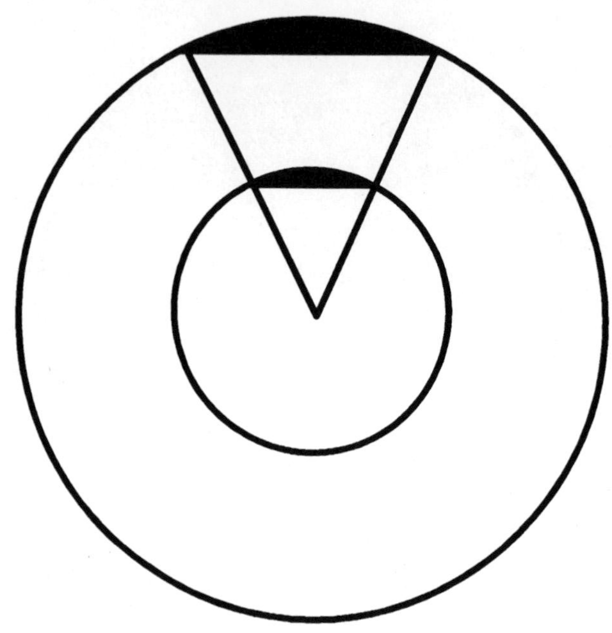

FIGURE 27. Two segments of circles are similar if they are formed by the same or equal angles.

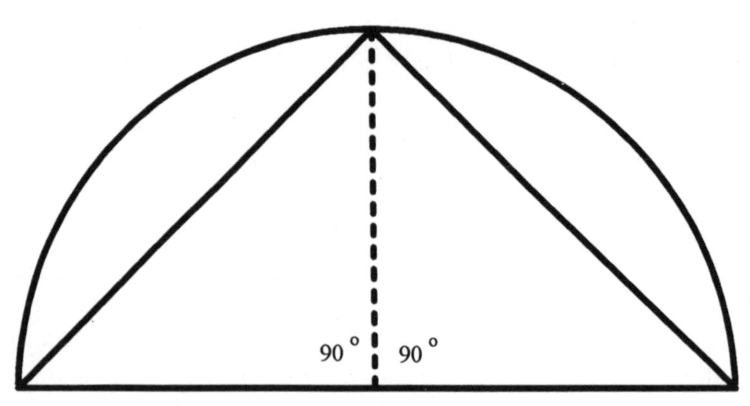

FIGURE 28. An isosceles triangle inscribed inside of a half-circle, creating two similar segments, both formed by a right angle.

Hippocrates now formed a segment with the base that was similar to the two smaller segments. How could he be sure that this larger segment was really similar to the smaller ones? It is easy once we recognize that the forming angle to both small segments is a 90 degree angle. We see this by dropping a perpendicular line down from the apex of our triangle to intersect the diameter. Each of the small segments can now be seen to be formed by a right angle.

We can now easily draw a segment along the diameter which has a forming angle of 90 degrees. In Figure 29 we have completed the semicircle and inscribed a second right triangle in the second half. The angle β we know to be a right angle of 90 degrees, and therefore its resulting segment is similar to the two small segments since all the segments are formed from 90 degree angles.

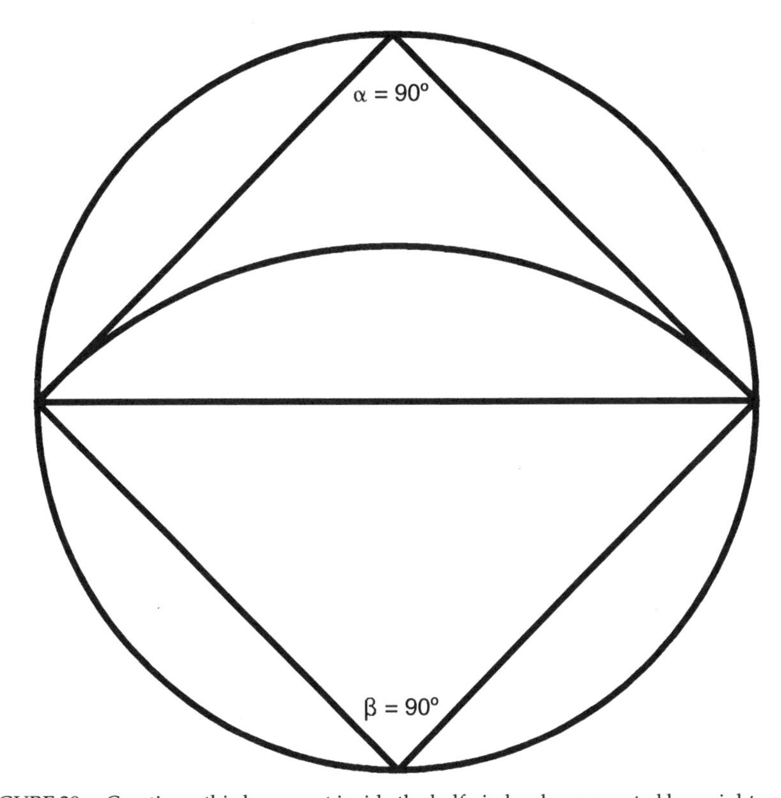

FIGURE 29. Creating a third segment inside the half-circle, also generated by a right angle.

The proof is all but complete. Since the ratio of the areas of similar segments is equal to the ratio of the squares of their forming lines, we see that:

$$\frac{\textit{Area Small Lune}}{\textit{Area Large Lune}} = \frac{(\textit{leg})^2}{(\textit{hypotenuse})^2}$$

Let h stand for the length of the hypotenuse and L stand for the length of each leg. From the Pythagorean theorem we know the hypotenuse squared of a right triangle is equal to the sum of the square of the legs, or $h^2 = L^2 + L^2$ or $h^2 = 2L^2$. In the above proportion, we can substitute $2L^2$ for h^2 and get:

$$\frac{\textit{Area Small Lune}}{\textit{Area Large Lune}} = \frac{L^2}{2L^2}$$

Therefore, the large lune is exactly twice the area of one of the small lunes. Hence, the area of the two small lunes together is equal to the area of the large lune. If we start with the area of the semicircle and subtract the area of the large lune, we are, in effect, subtracting the areas of the two small lunes. This demonstrates that the area trapped between the large lune and the semicircle is equal in area to the right triangle (Figure 30).

Thus was the proof of Hippocrates as reported by Simplicius through Eudemus. It is the first known case where an area bounded strictly by curved lines (the lune) was proven to be equal to an area bounded by straight lines (a right triangle). The proof led the Greeks to believe that it must be possible to square the circle, which meant finding some polygon

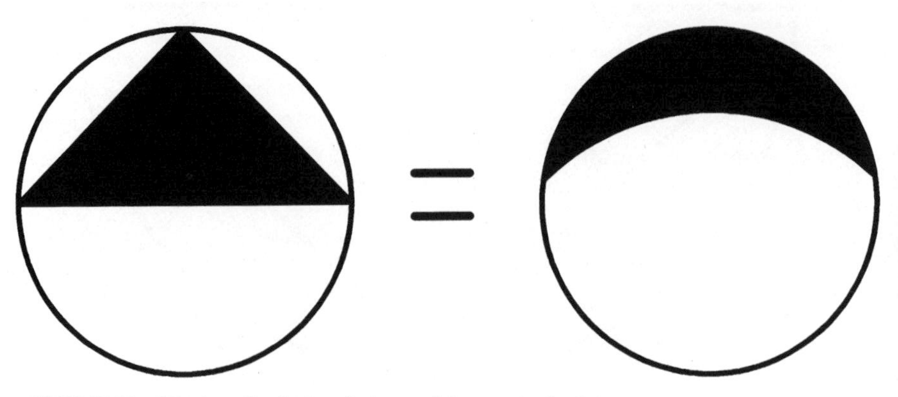

FIGURE 30. The inscribed triangle is equal in area to the lune.

(namely a square) equal to the area of a circle based on the radius of the circle. Yet, the proof had an even greater influence on mathematics, for the success of the quadrature of the lune encouraged mathematicians through the ages to attempt to get control of areas under curves. Eventually, this helped lead the way to modern calculus.

AXIOMATIC SYSTEMS

Up to this point we have talked as if proofs constituted an isolated feature or technique of mathematics which we could apply to specific cases in order to demonstrate a mathematical truth. However, an entire field of mathematics, known as axiomatic theory, relies on the idea of proof as its central theme. In axiomatic theory we begin with a set of axioms or postulates (the names are used almost interchangeably in modern mathematics). These axioms are statements about certain mathematical objects that are, themselves, left undefined. Such undefined objects we call primitive notions. All the characteristics of our primitive notions are supposed to flow from the axioms that describe them. From the axioms dealing with our primitive notions we deduce (prove) a long chain of theorems or conclusions. Once we have built such an axiomatic system we begin to feel secure that our theorems are consistent, and that if our axioms are not contradictory, then we will never be able to deduce two theorems that contradict each other.

Why should we want a system of consistent axioms and theorems? If we can find an interpretation within the real world of the primitive notions dealt with by the system, then we believe that we can then also interpret the theorems to be truths about the real world. The first organized attempt at this which has survived to the present day is the system of geometry produced by Euclid during the fourth century B.C. Euclid wrote *The Elements* in thirteen books. In the beginning he lists a total of five common notions and five postulates, all of which we would consider either axioms or postulates now. His five common notions deal with the relationship of equality:

1. Equals added to equals are equal. (We can add the same thing to both sides of an equal sign without loosing equality.)
2. Things equal to the same are equal to each other. (If $A = B$ and $A = C$, then $B = C$.)

3. Equals subtracted from equals are equal. (We can subtract the same thing from both sides of an equal sign.)
4. Those that coincide are equal. (Geometric figures that coincide are equal figures.)
5. The whole is greater than the part.

Euclid's postulates dealt with the notions of points and lines:

1. Between any two points a straight line can be drawn.
2. Every line can be extended indefinitely.
3. For every center and radius there exists a circle.
4. All right angles are equal to each other.
5. For every line and a point not on that line, there exists one and only one line passing through the point parallel to the given line.

From these axioms and additional definitions, Euclid was able to construct a beautiful system of geometry which stood as the superb example of an axiomatic system for two thousand years. His primitive notions were points and lines, yet he meant them to have the common interpretation of points in space and lines in space.

Until the nineteenth century Euclid's system stood alone as a meaningful axiomatic system. Finally, in 1889, the Italian mathematician Giuseppe Peano constructed an axiomatic system which dealt with arithmetic instead of geometry. Peano used only five axioms and two primitive notions. To fully appreciate that it is the axioms which define the primitive notions, we present his axioms using entirely neutral terms for the notions.

1. Heinsforth is a gelb.
2. If x is a gelb, the ranker of x is also a gelb.
3. Heinsforth is not the ranker of any gelb.
4. If two gelbs have equal rankers, they are equal.
5. If a set of gelbs contains the gelb Heinsforth and it contains all the rankers of its members, then the set contains all the gelbs.

The above axioms certainly look peculiar, and it is not obvious what gelbs and rankers, or even Heinsforth, are. But that is just the point. We don't want to give these primitive notions any names you would recognize, because it is only the relationships between the primitive notions captured by the axioms that matters. From this point we can go on and prove hundreds of theorems about gelbs and the relationship we have identified as rankers. Most will appear as meaningless as the above axioms. However, we have built a logical model awaiting some interesting interpretation.

Our next chore is to see if we can find mathematical objects which we can identify as gelbs and rankers which will obey the five axioms. If we can, then we will have an interpretation for our model. However, we must keep in mind that it will be only one interpretation, and there may well exist an infinite number of other possible interpretations. How about if we replace gelb with number and use ranker for successor. We shall also identify Heinsforth as the specific number one.

1. One is a number.
2. If x is a number, the successor of x is also a number.
3. One is not the successor of any number.
4. If two numbers have equal successors, they are equal.
5. If a set of numbers contains the number 1 and it contains all the successors of its members, then the set contains all the numbers.

For this system, successor and number are considered as primitive notions. It is easy to see that this interpretation of our axiomatic system yields the counting number sequence; 1, 2, 3, 4, . . . If we begin with the number 1, and define the idea of a successor of a number, then we can define the entire sequence of counting numbers by adding the number 1 to form each number's successor. Thus 2 is the successor of 1, or $2 = 1 + 1$.

Therefore, the various theorems that Peano proved regarding his primitive "number" turn out to be true about our counting numbers. However, there may exist other objects which could be interpreted as gelbs and other relationships as ranker and Peano's theorems would be true for them, assuming that his axiomatic system is an appropriate model.

REVISITING THE NUMBER LINE

In the previous chapter we talked at some length of the difficulties inherent in our notion of a common line, whether it could be made up of a finite number of points with magnitude or an infinite number of points without magnitude. We can use modern set theory to construct an axiomatic system that will help us understand the nature of the line. Such a system is called linear point set theory. We will pursue our study of the line by considering it as simply the infinite collection of a set of points. These points will have a certain relationship to one another, as stated in a set of axioms. Hence, we are going to build a system using a method remarkably similar to that employed by both Euclid and Peano. This new approach is the result of the work of Georg Cantor, who was able during the second half of the nineteenth century to show that it makes sense to talk about infinite sets which are of different sizes. Modern linear point set theory is now considered to be the foundation of all mathematics.

Set theory demonstrates all the remarkable relationships holding between points on a line. However, the theory is so broad that we don't have to think of it applying to lines and points only. Linear point set theory is about the relationship between objects that share certain characteristics. Any collection of objects sharing these characteristics will also share the demonstrated relationships. Therefore, we should think of linear point set theory as a general mathematical theory which can be interpreted as applying to points on a line, but that is only one possible interpretation.

This illustrates an important idea central to mathematics. It is possible to construct mathematical systems which talk about objects that are, themselves, left undefined. This is because the mathematician is not interested in the objects, but only in how the objects are related to one another. If the theoretical mathematician succeeds in constructing a logical system, the applied mathematician can then try to use this system to model some specific set of objects. The system will prove useful if it can be used to model a wide variety of different situations, either in the mental sphere or in the physical universe. Such is the case with linear point set theory and its application to points in order to model the mathematical line.

The Axioms of Linear Point Set Theory

Axiom 1: S is a collection of points such that

a) If x and y are different points, then either x precedes y or y precedes x. In symbolic notation we write: $x < y$ or $y < x$, using the symbol "<" to stand for "precedes."

b) If the point x precedes the point y, then x is not the same point as y ($x \neq y$).

c) If x, y, and z are points such that x precedes y and y precedes z then x precedes z (if $x < y$ and $y < z$, then $x < z$).

Axiom 2: S contains no point that precedes every other point (is a first point), nor does S contain a point that is preceded by every other point (is a last point).

Axiom 3: S is not the sum of two mutually separated point sets. (S is connected.) Considering the real number line, this axiom says that we cannot find a gap or hole on the line which is devoid of any numbers. This insures that the number line is a true continuum.

Axiom 4: S contains a countable set which is everywhere dense in S. (S is separable.) To be dense means that for every two points, $x < y$, there must exist a point, z, that is between x and y, or $x < z$ and $z < y$.

In this axiomatic system we leave the idea of point and the relationship of precedes ($<$) as primitive notions. We write the sentence "the point x precedes the point y" as $x < y$. However, you see at once that one interpretation of these notions is just what their names imply—"point" means a dimensionless position on the number line and the idea "precedes" means that our points are ordered in relation to one another. Notice that the axioms do not mention a line. The axioms are meant to define for us just what a line is, based upon the primitive notions of point and the relationship precedes. Hence, the theorems proved from this axiomatic system will help describe just those characteristics possessed by a number line.

Axiom 1 gives us the ordering relationship between points: namely, that if two points are not the same, then they share in the relationship we call "precedes." This makes intuitive sense because two different points on a line are such that one always precedes, i.e., is to the left of, the other. The other two parts of Axiom 1 state that if a point precedes another it cannot be equal to that other point. Axiom 2 states that there are no end points to our line; that is, no point exists which precedes all the other points, and no point is preceded by all other points.

Axiom 3 states that there are no gaps or holes in our line. If we consider just the rational numbers on the real number line, we can easily separate them into two mutually exclusive point sets. For example, π is not a rational number. We can separate the rational numbers into one set of all rational numbers that precede π, and a second set of all rational numbers that are preceded by π. Therefore, the set of all rational numbers is not connected, since it contains holes, namely all the irrational numbers. The set of all real numbers on the number line is connected, since we cannot separate them into two mutually exclusive sets.

Axiom 4 is a little more complex. It states that the points are dense everywhere along our line—there are no places where the points thin out. This is a rather sophisticated idea, yet it is this very characteristic of lines which gives them the strength to carry the great weight of responsibility we assign to them in mathematics. Specifically, Axiom 4 states that there exists a countable set of points which is a subset of S, such that every region of S, no matter how small, contains a member of the countable set. This means that no matter how small a region we choose, we can always find one point of the subset within that region. Considering the real number line again, the rational numbers on the number line are a countable set. At the same time, the rational numbers are dense everywhere on the line, i.e., there are no regions on the line that contain only irrational numbers.

The system represented by our axioms is incomplete because, for brevity, I have not provided you with all the definitions of the terms within the axioms (e.g., mutually separable, connected, everywhere dense) based on the notions of point and precedes, which would be required in any formal presentation. However, we should consider the definition within our axiomatic system of the idea of a limit point, since it is of critical importance to calculus.

DEFINITION: A point P is said to be a *limit point* of a point set M if every region containing P contains a point of M distinct from P.

Let's say that P is the president of the United States, while the point set M represents all the Secret Service agents assigned to protect the president. With a finite number of Secret Service agents, we realize that one of them must be standing closest to the president. Hence, there would exist some small region around the president that contained no agent. In this case, the president would not be a limit point to the set of Secret Service agents. Now pretend the president has an infinite number of Secret Service agents. He positions one at a hundred yards' distance, the second at fifty yards, the third at twenty-five yards, and so forth, until all are positioned, each one closer to him. For this to work, we must pretend the agents diminish in size. Now we say that the president is the limit point, P, for the set of Secret Service agents, M. Any small circle we draw around the president must

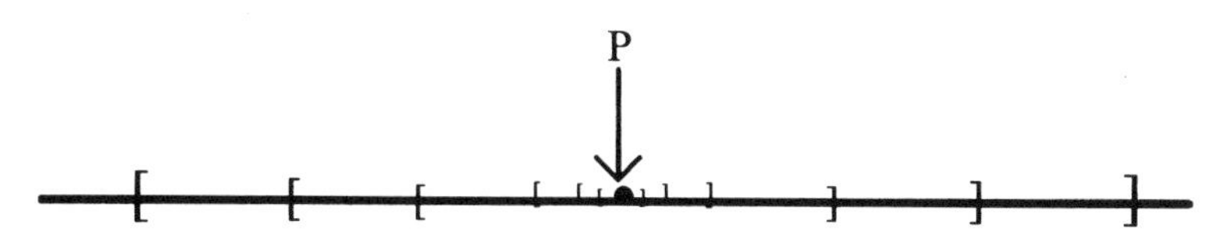

FIGURE 31. A number of regions on a number line all containing the limit point, P. P is a limit point for the set M if every conceivable region that contains P also contains some point of M.

contain at least one Secret Service agent. In fact, it will actually contain an infinite number of them.

Figure 31 shows a number of regions of a line all containing our limit point, P. To say that P is a limit point for the set M is to say that every conceivable region that contains P also contains some point of M. For this to hold, we must visualize that some subset of the points of M must continually draw closer to the point P.

A SET THEORY PROOF

Now let's try proving a theorem from our axiomatic system.

Theorem: If x is a point of S, then there exist two points, A and B, such that A precedes x and x precedes B. We call all the points between A and B the region AB, and we say the region AB contains x.

Proof: Step 1: Assume that there exists a point x such that no region contains x. This is a negation of the theorem and we are hoping to prove a contradiction when we assume it is true. If we can show such a contradiction, then we will know that the opposite must be true—and the opposite is that our original theorem must be true.

For example, another proof using contradiction was our proof that the counting numbers continue without end. We assumed there existed a largest counting number. This assumption was the negation of our original conjecture. Then we showed that this assumption led to a contradiction by simply constructing a larger number, i.e., by adding one to it. If the negation of our original conjecture leads to a contradiction, then the original conjecture (that the counting numbers continue without end) must be true.

Step 2: If the assumption in Step 1 is true, then there exists no point A such that $A < x$ or there exists no point B such that $x < B$.

Step 3: Assume that there exists no A where $A < x$. If this is true then x is a first point of S, which is a contradiction to Axiom 2, which states that S has no first or last point. Hence, we know that there must exist a point A where $A < x$.

Step 4: Assume that there exists no B such that $x < B$. If this is true then x is a last point of S, which is also a contradiction to Axiom 2. Hence, there must exist some point B such that $x < B$.

Step 5: Since an A and B exist such that $A < x$ and $x < B$ then AB is the region containing x. Theorem proven.

The above example was a small proof. Many theorems have proofs which ramble on line after line. One fun challenge for a mathematician is to take an existing proof for a theorem and reduce it in size and complexity. This is considered providing a more *elegant* proof, for mathematicians love simplicity.

Figure 32 shows the logical relationship between the various parts of an axiomatic system. First we have those items which are a given: the axioms, the primitive notions, definitions, and rules of inference. The definitions do not really add anything to the system but are used to

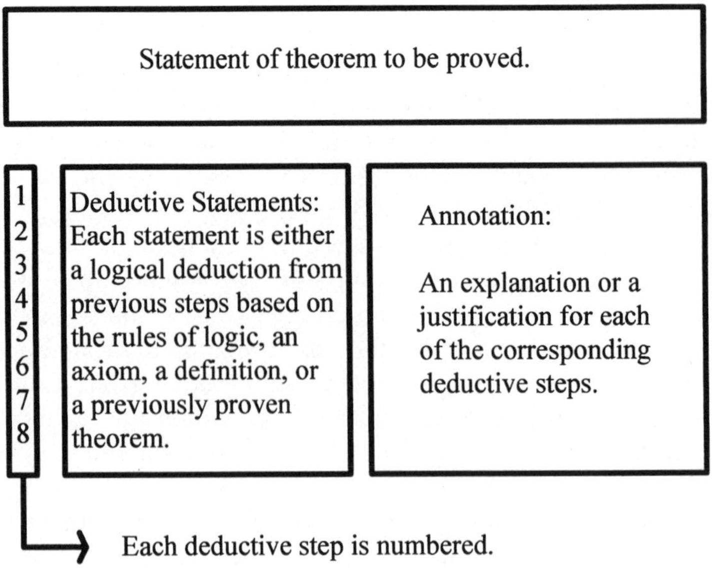

Statement of theorem to be proved.

1
2
3
4
5
6
7
8

Deductive Statements:
Each statement is either a logical deduction from previous steps based on the rules of logic, an axiom, a definition, or a previously proven theorem.

Annotation:

An explanation or a justification for each of the corresponding deductive steps.

Each deductive step is numbered.

FIGURE 32. Diagram of the parts of a mathematical proof.

simplify statements so that they can be more easily used later on. Next come the theorems. Each theorem must be proved using only those items in the box above (especially the axioms), or theorems which have been proven before. These will appear on our list of theorems before the theorem we are currently proving.

The third box shows the body of the proof for a specific theorem. Each line of the proof will contain a statement (proposition) in symbolic algebra which must be justified by a corresponding rule of logic. Every theorem on the list must have its appropriate proof in order to be included in the list. The layout for any specific proof varies according to the tastes of individual mathematicians, but one popular format is to write the symbolic statement (proposition) on the left and the justification for the statement (annotation) on the right.

To catch the flavor of a proof as written by a working mathematician, we should become familiar with several symbols from symbolic logic. Such symbols are really secondary to the proof and are used as a shorthand to make the written proof move faster and to remove any ambiguity found in English.

$\forall$	"for every"
$\exists$	"there exists"
$\in$	"is a member of"
$\neg$	"negation"
$\mid$	"such that"
$\wedge$	"and"
$\vee$	"or"

Now we are ready to give a symbolic rendition of our proof.

	Proposition	Annotation
1.	$\exists x \mid \nexists AB \mid x \in AB$	Assumed negation of theorem
2.	$\forall A \mid A \neq x, (x < A) \vee \forall B \mid B \neq x, (B < x)$	Definition of region, Step 1
3.	$\forall A \mid A \neq x \ (x < A)$	Assumed from Step 2
4.	x = First Point	Definition of a first point
5.	Contradiction, step 4	Axiom 2
6.	$\forall B \mid B \neq x \ (B < x)$	$(p \vee q) \wedge (\neg p) \supset q$
7.	x = Last Point	Definition of a last point
8.	Contradiction, step 7	Axiom 2
9.	$\neg[\exists x \mid \nexists AB \mid x \in AB]$	Negation of contradiction
10.	$\forall x \ \exists AB \mid x \in AB$	Equivalence to Step 9

By using symbolic logic we have tried to avoid the ambiguity found in everyday language. This proof followed the general course of our first proof. That is, we assumed the theorem we wished to prove to be false. From this we demonstrated a contradiction which meant our theorem was, in fact, true after all.

When we build modern axiomatic systems we wish them to have two characteristics. First we want them to be complete. For a system to be complete implies that any statement written in terms of the primitive notions can be proved to be either true or false. Some small axiomatic systems are known to be complete. Second, we want our system to be consistent. This means that we will never prove a theorem which turns out to be a contradiction to one of our axioms or other theorems. This would be a disaster, since once we have proved one contradiction, we can use it to prove any and every statement within the axiomatic system regardless of whether it is true or not. Hence, consistency is very important.

Suddenly enters a young Austrian mathematician by the name of Kurt Gödel (1906–1978). In 1931 Gödel proved that for an axiomatic system substantial enough to cover arithmetic, we cannot demonstrate that the system is both consistent and complete. This is known as Gödel's Incompleteness Theorem. Thus Gödel demonstrated that there will always be true statements about the positive integers which we cannot prove with deductive reasoning.

When he presented his proof, it turned mathematicians on their heads. Many had been attempting to build a system which would be both consistent and complete for all of mathematics. Now, they realized it was an impossibility. This means that for any axiomatic system (using our current system of symbolic logic) large enough to be really useful, we must give up hope of completeness. This implies that there will always be true statements within the system which we cannot prove to be true. Now we have to do without completeness, since consistency is an absolute necessity. The need for consistency explains why mathematicians want to develop axiomatic systems that have as few axioms as possible. If we just throw statements willy-nilly into our set of axioms, there is a danger we will, unwittingly, include a statement as an axiom which contradicts one of the other axioms. In order to avoid this, we try to keep the number of axioms as small as possible.

We have provided examples of proofs from ancient Greece and our modern era. Yet, in all these centuries the idea of a proof stayed the same—

a demonstration to convince the reader of the truth of the conclusion. We cannot overstate our indebtedness to the Greeks for introducing the idea of proof into humankind's search for truth. While other ancient societies were hampered for centuries, even millennia, by tradition and authoritative doctrine, the ancient Greeks broke free by demonstrating that truth can stand on its own legs, and need not be propped up by authoritarianism. The idea of proof has spread beyond mathematics to science in general and helps give science its credibility. If we were to look for a distinguishing characteristic between mathematics and science on one hand, and other fields of endeavor, such as occult science or religion on the other, we would see that the idea of demonstration is critical to the former and not the latter. A cornerstone of religion has always been faith. Faith is belief as opposed to knowledge. When something is proved to us, it becomes knowledge and there is no reason to rely on faith. As George Simmons, author and professor of mathematics, so elegantly states:

> Thus skepticism was at the heart of the Greek way of doing philosophy, which is unique in the history of thought; it led to the concept of mathematical proof—which arose nowhere else—because skepticism and demonstrative proof are opposite sides of the same coin; and this notion of proof is the essence of genuine mathematics, which gave rise in turn to its companion, mathematical science, both of which also appeared nowhere else.[7]

Americans, of all peoples, should be quick to embrace mathematics, for mathematical proof goes hand-in-hand with democracy. Mathematical proof is free of class, special privilege, and the governing elite. It is open to all, regardless of station, and only requires a skeptical mind and a willingness to engage in rational thought.

4

Passing the Torch

After the decline of the Greek city–states, the Roman Empire dominated the entire Mediterranean area. The Romans were less known as philosophers and better known as great administrators and empire builders. They didn't seem to relish the pure joy of doing mathematics or philosophy just for the pleasure of the act. While they admired what the Greeks had achieved, they were too busy fighting wars and building roads to be bothered by purely intellectual pursuits.

Therefore the spotlight for mathematical discovery shifted away from Europe and toward Africa and Asia. The ancient Chinese civilization was very old, going back as far as the reign of Fuh-hi (2852–2738 B.C.), the first emperor of China. During his reign, the Chinese were making extensive observations of the positions of stars and beginning to evolve a strong astronomical tradition. Because of the isolation of China, westerners tend to ignore their contributions, while concentrating on the achievements of others. Overall, the Chinese were more advanced than either the Babylonians or the Egyptians, but still less sophisticated than the Greeks. Sometime around or before 1000 B.C. the Chinese began using both red and black sticks on their counting boards, the red sticks representing positive numbers and the black ones representing negative numbers. Over the centuries their use of negative numbers may have spread as far west as the Hindu culture in India. In any case the Chinese generalized numbers beyond positive integers and positive fractions to include negative rational numbers. At first, the adoption of negative numbers was not complete, since they did not allow negative numbers to stand for solutions to equations.

The next instance of generalization within mathematics came from the Hindu society of India which began around 1500 B.C. when the Aryans migrated south into India. No mathematical manuscripts have survived from the first millennium, but other works indicate that mathematics was being pursued. From 500 B.C. on we do find a recorded tradition of Hindu mathematics until the time of Brahmagupta (ca. A.D. 628), who worked in numerous areas including interest rates, geometric areas and volumes, and algebra. He wrote the first comprehensive work on handling both negative numbers and zero. This greatly expanded the mathematical objects considered as numbers beyond the Greeks' positive integers and their ratios.

Hence, it was really the Asian contribution, through both the Chinese and the Hindus, that completed the set of numbers we call rational numbers, e.g., all numbers of the form p/q where p and q can be any positive or negative integer (with the condition that q is never zero). This greatly benefitted algebra for it meant that certain simple equations now had meaningful solutions. For example, the equation $x + 7 = 4$ had no solution for the Greeks, but for Brahmagupta the solution was easy; it was just -3, or minus three.

The work of Brahmagupta came after the end of the ancient Greek period of productivity and many centuries before the European Renaissance. At the beginning of the European mathematical Renaissance, negative numbers first appeared in a work on arithmetic and algebra by Nicolas Chuquet in 1484. However, centuries would pass before all Europeans fully accepted both the negative numbers and zero as legitimate objects of mathematics. Yet, some mathematicians were not going to wait until everyone was on board. They were ready to discover modern mathematics.

THE BEAUTY OF EQUATIONS

We are now ready to consider analytical geometry—the contribution by two great seventeenth-century mathematicians, René Descartes and Pierre de Fermat. Yet, before we do, we must take a short detour to talk about equations, for much of what we want to introduce concerns equations. For many math phobics, equations are especially loathsome since they may make the heart race and the palms sweat. "Why can't we just say in words what is meant in mathematics and leave all those disgusting equations out?" But we really can't do that. To fully understand and enjoy mathematics we must come to appreciate equations. Mathematics is about relationships, and many relationships are best demonstrated with the language of symbolic algebra. To understand mathematics without equations is like understanding great art without looking at paintings—impossible!

The solution is simple. We must come to look at an equation as we would a great work of art. We study it, and come to comprehend and appreciate it. Many of our equations are very simple, yet harbor deep truths. Others are very elaborate and elegant, and fun to analyze, giving the eye a pleasant treat and the mind a startling insight.

Mathematics has frequently been compared to music, and the similarities are rather intriguing. Both mathematics and music are linear in that both depend upon a sequence of events, instead of everything happening at once, as we experience in the visual field. In music we hear notes in sequence, through time. With mathematics, we count with numbers, through time; and we work a problem or a proof with a sequence of steps, again through time.

One of the greatest joys we have with music is our sense of anticipation. Once we learn a piece of music, we find we love to hear it over and over, because our minds anticipate the next note or set of notes. Then when we hear these notes, we experience the pleasure of completion. Great music is music we enjoy as we hear it for the first time because the melody is one we immediately catch and begin to anticipate. Many great classical themes are simple and presented within the work over and over again, allowing us to immediately anticipate and enjoy. Mathematics also pleases us through anticipation. The very basic act of counting is to anticipate the next integer in the sequence. As we learn more sophisticated mathematics we encounter infinite sequences and series, and we see from their form what the next unlisted term will be—giving us pleasure in the fulfillment of our anticipation. Mathematical problem solving and proofs also allow us to anticipate and then feel pleasure in the completion of that anticipation.

When teaching mathematics I must repeat many mathematical demonstrations on the chalkboard before my classes. Yet, the demonstrations never grow boring, because I always get that little jolt of pleasure from completing the act of demonstration. It's like listening to good music. Some days, when asked by a student to demonstrate a difficult problem, I tell myself that I shall only do the beginning of the demonstration and leave the conclusion for the student to complete. Yet, once I begin the dance of the chalk upon the chalkboard I become so lost in pleasure that before I can stop myself, I've finished the entire problem.

Mathematics and music allow for complexity. When things get too simple and repetitive, we can add new themes to both music and mathematics. Some classical music is so complex that we can listen to it for years and still hear new nuances. Complex mathematical equations can contain hidden surprises in a similar fashion.

In both fine music and mathematics we sense perfection. With music we often feel that the piece as presented is perfect. To change one note would ruin it. Of course, when a talented musician comes along and does change it, we suddenly realize the original was not perfect, but that the new version is now precisely perfect! Mathematics is the same. The equal sign in the equation means exactly equal. If the two sides are not exactly equal, then we use a different sign. Hence, when we see two entirely different forms (one on each side of the equal sign) and realize they are truly equal, we are astounded at the perfection of this equality.

The last shared characteristic I offer between mathematics and music is the potential for related themes. In music we may have different themes, but when they are played together, the result is harmonious and sounds better to the ear than any one theme played alone. A symphony orchestra playing a rich classical piece has many different themes played at once, and the result can move an audience to tears. Mathematics also weaves different themes together into a rich tapestry. While working in one field of mathematics we suddenly realize that what we are doing is directly applicable to a completely different field in a unique and surprising way. This realization gives the mathematician a sudden rush of profound pleasure. It is like a narcotic; once experienced, you can never get enough.

Think of a moving piece of music that stirs your emotions. You're listening to the radio and this piece suddenly comes on. You immediately turn up the volume, and are instantly transported to some magical world of beauty and pleasure. The same happens to a mathematician when confronted with a particularly intricate and deep mathematical expression. We all have our favorites and I offer you one of mine. Consider the following equation (which really relates three things to each other) as you would a work of art, and let's see if we can discover the magic within.

Infinite Nested Radical	Golden Mean	Infinite Continued Fraction

$$\sqrt{1+\sqrt{1+\sqrt{1+\sqrt{1+\ldots}}}} = \frac{\sqrt{5}+1}{2} = 1+\cfrac{1}{1+\cfrac{1}{1+\cfrac{1}{1+\cfrac{1}{1+\ldots}}}}$$

The above equation contains three different ways to write the exact value of the golden mean, ϕ, discovered by the Greeks. On the left is the simplest infinite nested radical, in the middle is a simple finite algebraic expression for ϕ, and on the right is the simplest infinite continued fraction. Whenever I see this expression a tingle goes up my back. How can the simplest infinite continued fraction be exactly equal to the simplest infinite nested radical, and how can they both be equal to the Golden Mean? Finally, these forms are pleasing to the eye, for they possess a delicate pattern.

Now that we've appreciated the beauty of an equation, we will not be repulsed in displeasure when we encounter the next one. Instead, we are going to feast our eyes upon it, learning its deep secrets and relishing its form.

THE REMARKABLE DESCARTES

We must now turn our attention to one of the most remarkable men who ever lived, René Descartes (1596–1650). Not only was Descartes one of the most gifted mathematicians for his own time (only Pierre de Fermat rivaled him), but he was an outstanding philosopher as well (Figure 33). His life's story is so rich and varied, that it seems incredible that he did so much and still died before reaching the relatively young age of fifty-four. Descartes was born of noble parents on March 31, 1596 in Touraine, France. Within a few days of his birth, his mother died, and the doctors proclaimed that the sickly child, René, would soon die also.

Yet he lived, but was never a healthy child nor a robust adult. At the tender age of eight he entered the Jesuit college of La Flèche in Anjou, France where he studied the classics for the next eight years. In 1612 he went to Paris where he entered the University of Poitiers, earning a law degree in 1616. But law was not to be his great love; that was reserved for mathematics and philosophy. At the age of twenty he was not ready to settle down, so he entered a military school at Breda, Holland. A turning point occurred in his life on November 10, 1618 while stationed in Breda. He observed a crowd gathered on the street before a notice. Asking a bystander to translate the Flemish, he discovered that the notice was a mathematical problem and an invitation to solve it. In the presence of the crowd, Descartes remarked offhandedly that the problem was easy! The bystander, who turned out to be the head of the Dutch College at Dort, Isaac Beeckman, challenged Descartes to make good his boast and find the

FIGURE 33. René Descartes (1596–1650). Photograph from Brown Brothers, Sterling, PA.

solution. Descartes did, and Beeckman at once recognized the great talent possessed by the young soldier. Beeckman encouraged Descartes to continue his mathematical studies, presenting him with several worthy problems to solve.[2] Two years later Descartes was still in Holland studying science under the tutelage of Beeckman.

However the wanderlust continued to possess Descartes, and in 1619 he joined the Bavarian army. For the next nine years Descartes traveled about Europe and fought with several armies, including those of Maurice, Prince of Nassau, Duke Maximillian I of Bavaria, and in the French army at the siege of La Rochelle. During this time Descartes, like many young men from noble families, developed a taste for women and gambling (he is reported to have been a winner).[3] Yet, his mathematical and philosophical ideas never left him. He used his military travels as an opportunity to meet a variety of men of science throughout Europe.

One notable acquaintance was the Minimite friar, Father Marin Mersenne (1588–1648). Mersenne had attended the same Jesuit College at La Flèche as Descartes. He lived at a monastery near the Palace Royale in Paris and held regular meetings at the monastery for scientists and mathematicians. These meetings continued even after Mersenne's death and eventually evolved into the French Academy of Sciences in 1666.[4] In addition to the meetings, Father Mersenne acted as a conduit for the flow of mathematical ideas between various European mathematicians, including Descartes, Galileo, Torricelli, Cavalieri, Fermat, Hobbes, and many others. Frequently new ideas in mathematics were channeled through Mersenne to other mathematicians before they were published. In some cases, this resulted in arguments regarding priority of discovery, which we will encounter between both Descartes and Fermat, and between Leibniz and Newton. In addition to acting as a clearing house for mathematical ideas, Mersenne contributed significantly to the study of sound, earning the title "Father of Acoustics."

In 1628, at the age of thirty-two, Descartes decided to cease his wanderings and settle into one place. He chose Holland because that country seemed to have an especially liberal attitude for innovative thinking. Descartes' first major work was written at this time: *Le Monde, ou Traité de la Lumiére*, a work on physics. Unfortunately, an event in Italy caused him to hold this work back from publication for fear of retribution from the Catholic Church.

The event centered around Galileo Galilei (1564–1642), an older contemporary of Descartes. Galileo (Figure 34) was famous in Europe for his many inventions and his study of motion. However, he had clearly positioned himself on the side of the Copernicans, who claimed that the earth and other planets revolved around the sun, instead of the Aristotelian and Catholic Church doctrine which said the Sun, planets, and stars all revolved around the Earth. In 1632 Galileo published his exquisitely argued *Dialogue on the Great World Systems*, demonstrating the superiority of a heliocentric theory of the solar system. It did not agree with church leaders, including the pope, and Galileo was summoned to Rome where

FIGURE 34. Galileo Galilei (1564–1642). Photograph from Brown Brothers, Sterling, PA.

he was tried and convicted on a charge of heresy. He was forced to recant his heliocentric theory, after which he was sentenced to house arrest and forbidden to publish for the remainder of his life. Galileo's various works covered the spectrum from theoretical mathematics to improving the telescope and discovering the moons of Jupiter. He contributed to the theory of hydrostatics, the motion of fluids, with a theory for both floating bodies and tides, and invented the pendulum clock and thermometer. He worked for many years to discover the path an object takes under the influence of gravity only to have the final discovery made by his student, Cavalieri (1598–1647), who demonstrated that (neglecting air resistance) an object thrust into the air follows the path of a parabola, one of Apollonius' conic curves.

News of Galileo's heresy trial and its outcome was not lost on Descartes. Intellectual thought in Europe, even Holland, was still dominated by the Catholic Church, and although Descartes was a resident of Holland, it would have been unwise for him to attack the Inquisition. His work on physics was not published until after his death.

In 1637 Descartes published his work on the methods of science, *Discours de la méthod pour bien conduire sa raison et chercher la vérité dans les sciences*. One of the three appendices to his work was *La Géométrie*, in which he outlined a method to connect the expressions of algebra with the diagrams of geometry. This new approach would forever change the face of mathematics. It combined both algebra and geometry under one specialty—analytic geometry. One story says that Descartes discovered his analytical geometry one morning while lying in bed watching a fly crawl on the ceiling. He realized that the path of the fly could be described using only the distances from two of the four walls.[5]

Remember that the ancient Greeks had discovered the diagonal of a unit square was incommensurable to the length of the sides. Since algebra dealt with numbers, which, for the Greeks, were limited to the positive integers, then there existed lengths in geometry (i.e., the diagonal) which could not be measured using integers. Hence, until the work of Descartes and Fermat, algebra and geometry were studied under different umbrellas. As we are about to see, the joining of algebra and geometry produced such a powerful form of mathematics that the door soon flew open for others to discover the higher mathematics of calculus.

In passing, we note that in 1637 Descartes published *La Géométrie* in Leiden, Holland; one year later, Galileo smuggled his work, *Discourses on Two New Sciences*, dealing with his work on motion, out of Italy and into Leiden, where it was published.

Descartes continued to produce important work. In 1641 he published *Meditations on First Philosophy*, which has come to be known in philosophy as simply, *The Meditations*. One legend is that he first drafted *The Meditations* while campaigning as a soldier. During a lull in the fighting he and his comrades relaxed in a Dutch bakery. His friends were drinking, gambling, and making a terrible ruckus. In order to find some quiet place to write, Descartes climbed into one of the large, unused ovens and closed the door. Inside, with candle, paper, and pen, he wrote the first draft to *The Meditations*. It is within this work that we find the oft quoted words: "I think, therefore I am." Descartes' contributions to philosophy cannot be overlooked. He was a modern rationalist who believed that we should believe only in claims to the degree justified by the evidence for such claims. This is opposed to the appeal to authority that was common for his day, i.e., if one wanted to know the truth, one simply consulted the experts, such as Aristotle or the pope. Descartes was willing to do away with the traditional authoritative test for truth and let each claim to truth live or die according to its factual and logical support. This was a very novel idea for its time.

In 1644 Descartes published his comprehensive work, *Principia Philosophiae*. Five years later at the invitation of Christina, Queen of Sweden, he moved to her court to give her instruction in philosophy and mathematics. Unfortunately, the rigorous regimen, beginning at five in the morning, combined with the colder northern winters, wrought havoc on Descartes' sensitive system. He died of pneumonia February 11, 1650, seven weeks before his fifty-fourth birthday.

FERMAT, THE AMATEUR

Pierre de Fermat (1601–1665) has been described by some as the greatest mathematician of the seventeenth century (Figure 35).[6] At the very least he is in company with the giants of mathematics: Archimedes, Newton, Euler, and Gauss. He contributed significantly in four areas: analytical geometry, calculus, probability, and number theory. Yet he was not a professional mathematician. He published only one work during his lifetime, leaving much of his work unknown until it was published by his son after Fermat's death. Even today, historians sometimes fail to recognize his contributions because they frequently occurred in the form of unpublished manuscripts circulated by Father Mersenne.

Fermat was born August 20, 1601 to Dominique and Claire Fermat in Beaumont-de-Lomagne (Tarn et Garonne), France. His father was a successful merchant while his mother was the daughter of a prominent French family.[7] Not much is known of his life simply because it was rather pedestrian, totally unlike that of Descartes. Before 1631 Fermat graduated with a degree in law from the University of Orléans. Soon after, he purchased (for 43,500 *livres*) a position of councilor to the local parliament of Toulouse and served on the provincial supreme court. He seems to have made little impact on law or politics for his time. Fortunately for us, he had sufficient spare time to devote to his hobby—mathematics. He mar-

FIGURE 35. Pierre de Fermat (1601–1665). Photograph from Mary Evans Picture Library, London, England.

ried his fourth cousin, Louise de Long and raised five children. After a career of unremarkable law and stupendous mathematics, Fermat died January 12, 1665, and was buried in Castres, southern France.

During his life, Fermat's mathematical works were known to others through his shared manuscripts with Father Mersenne. His eldest son, Clément–Samuel, published his father's works in 1670 and 1679. It is through his unpublished manuscripts that we know of his analytic geometry, work on areas and tangents (calculus), and his cofounding of probability. Yet, it was only after his death that his great work in number theory became public, for during his lifetime he could not interest any of the other mathematicians in this most fascinating field.

Today Fermat's name is most frequently associated with a particular number theory problem. In a story which has been told many times, he wrote a comment in the margin of one of his math books to the effect that, while a square of some numbers can be written as the sum of two other squares (e.g., $5^2 = 4^2 + 3^2$), this was not possible with higher powers. This can be written as the conjecture: $A^n = B^n + C^n$ has no integer solutions for A, B, and C when $n > 2$. Fermat went on to scribble in the margin that he had a nice little proof of this conjecture but insufficient space to include it. For the next 350 years mathematicians pulled out their hair, attempting to prove or disprove what became known as Fermat's Last Theorem.[8] Finally in 1994, Andrew Wiles of Princeton University produced a valid proof. Fermat's Last Theorem has been put to rest.

As cofounders of analytic geometry, Descartes and Fermat could not be further apart as individuals, even though they both possessed law degrees. Descartes had a dynamic career as both a mathematician and a philosopher. He was known throughout Europe, and produced numerous works during his lifetime. His personal life was full of danger and adventure since he was known for his gambling, soldiering, and luck with women. Fermat was the respectable opposite. He received his degree, married, and spent his quiet life raising children and practicing law. Fermat claimed to have invented his version of analytic geometry in 1629, which he described in a manuscript entitled *Introduction to Plane and Solid Loci*. A few months before Descartes published his work on analytic geometry, Fermat sent to Mersenne this very manuscript, which Mersenne then sent to others, including Descartes. Who discovered analytic geometry first? Most historians have given Descartes credit, for he published first. Yet, Fermat's manuscript was in the hands of others before that time. This question of priority caused considerable discomfort for Descartes who willingly vented his displeasure in public.

What was this outstanding contribution called analytic geometry? It was the combining of algebra with geometry. Fermat and Descartes were able to use a coordinate system to represent algebraic equations as geometric curves. What we now call analytical geometry started with these two men. This is not the entire system we use today, but it took their initial ideas to provide the basic framework. Not only did they combine algebra and geometry, but Descartes' handling of symbolic algebra was generally superior to his predecessors. His symbolic notation was not too different from the notation we use today. He used a backward proportional sign ($\propto$) for the equal sign (=), yet he used both the modern plus (+) and minus (−) signs along with modern exponential notation. He adopted the convention of using letters from the beginning of the alphabet (a, b, c) for constants and those from the end of the alphabet (x, y, z) for unknowns. Descartes helped shape symbolic algebra as we know it today. We cannot over-emphasize the importance of a clear and simple notation for algebra. Prior to Descartes, different mathematicians had used their own notational innovations. This caused difficulties for them when they tried to communicate with each other, or leave easily understandable work for future generations. Descartes' effort to formalize a standard notation greatly increased the sharing of mathematical knowledge from his time on.

Fermat was more of a geometrician, while Descartes was clearly the algebraist. Fermat went beyond Descartes in using two perpendicular axes in his coordinate system, which is called a rectilinear coordinate system. He found the rectilinear equations for lines, circles, and the three conic curves—parabola, hyperbola, and ellipse. Fermat discovered many new curves by simply writing down new algebraic equations and then investigating their corresponding graphs using his perpendicular axes.

THE WONDERS OF ANALYTIC GEOMETRY

Modern analytic geometry is based on a very simple idea: all locations in space can be given addresses! First we take two lines, one horizontal and one vertical, and cross them at right angles. Where the two lines intersect is called the origin. Next we mark off a scale on each line (Figure 36). At the origin we have zero for both lines. As we move along the horizontal line toward the right, we mark off the positive integers in order, and as we move toward the left we mark off the negative integers in order. This is just like our number line (see Figure 6), with all the spaces between integers filled with the other real numbers. We scale the vertical line in the same way with the positive numbers progressing upward and the negative numbers moving downward. Thus we have constructed a coordinate plane, called a Cartesian coordinate system in honor of Descartes. The two lines are called axes.

The remarkable feature of this plane is that if we pick any point in the plane, we can assign a unique pair of numbers to it which immediately identifies the point's location. Consider the point shown in Figure 36. We measure the distance along the horizontal axis and make this the first number in the pair. The second number is the distance along the vertical axis. Hence, the unique address of our point in Figure 36 is 4 and 3, generally written in parentheses as $(4, 3)$. A point need not have an address

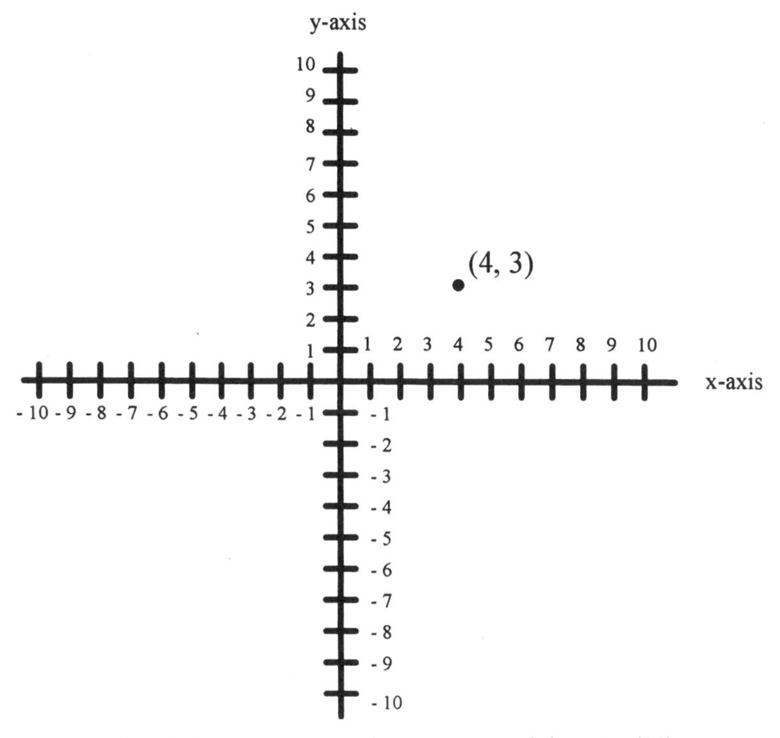

FIGURE 36. Cartesian coordinate system and the point (4,3).

involving integers, but can have any real number. Thus, we have a point corresponding to the address $(\pi, \sqrt{5})$, and $(-2, 0.412)$, etc. Any set of two real numbers gives us an address for a point and every point has a unique address. The addresses of points are called coordinates. For convenience, the first coordinate is sometimes called the x-coordinate (or abscissa) and the second is called the y-coordinate (or ordinate). Now it is possible to relate the points in geometry to the numbers in algebra. When we do this some very remarkable things begin to happen.

First, let's consider an equation with two unknowns, x and y, such that every term of the equation has no more than an x or a y or a constant. We called this a linear equation. Its most general form consists of three terms: $Ax + By = C$, where A, B, and C are just real numbers called coefficients. Notice that we don't have a term where the x and y are multiplied together and we don't have terms involving x and y raised to a higher power (exponents greater than 1). Now we get to the first wondrous example of the power of analytic geometry. Every linear equation when translated onto the Cartesian coordinate system of two axes turns out to be a straight line! That's right! Every linear equation is a straight line, and every straight line has a corresponding linear equation. Actually, every straight line is associated with an infinite number of linear equations. For example, the two equations $x + y = 1$ and $2x + 2y = 2$ are associated with the same straight line. They are dependent on each other because the second equation is simply the first equation with all the terms multiplied by the constant 2. This tells us that equations that are multiples of each other represent the same line in the Cartesian coordinate system.

Figure 37 shows a number of specific straight lines with their corresponding linear equations. Several kinds of equations are of special interest. A linear equation where the x-coefficient is zero gives us the linear form $By = C$, which is a horizontal line. Every linear equation where the

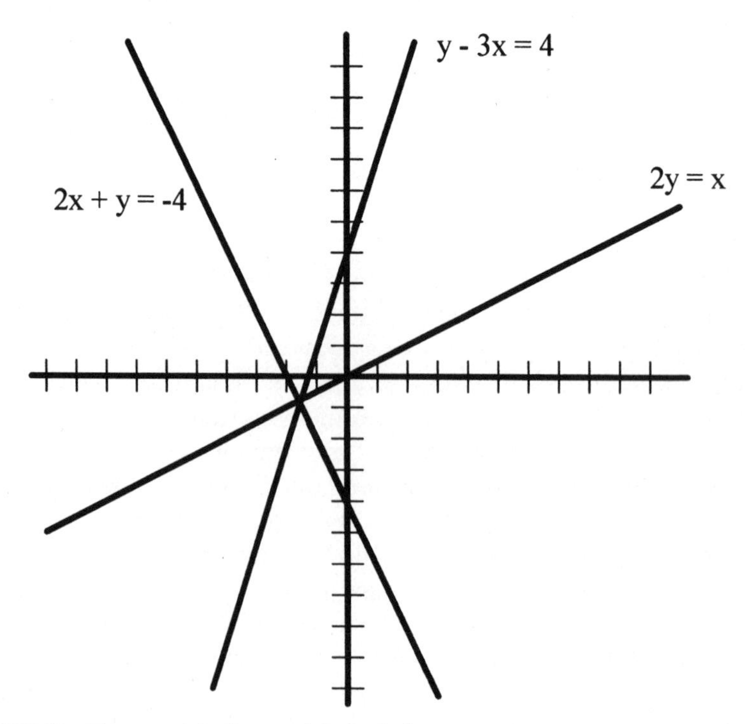

FIGURE 37. Three straight lines and their algebraic equations on a Cartesian coordinate system.

y-coefficient is zero, giving us $Ax = C$, is a vertical line. Every line that is neither horizontal nor vertical will have both an x and y term in its equation. If $C = 0$ then the line passes through the origin.

How do we take an equation and turn it into a line? It's very simple. Suppose we have the linear equation $4x - 2y = 3$. First we chose any number we want for x, say $x = 1$. Then we plug 1 into our equation to find out what the corresponding y is, or

$$\begin{aligned} 4(1) - 2y &= 3 && \text{substituting 1 for } x \\ 4 - 2y &= 3 && \text{multiplying } 4 \times 1 \\ -2y &= -1 && \text{moving the 4 to the right and subtracting} \\ y &= 1/2 && \text{dividing by } -2 \text{ and simplifying.} \end{aligned}$$

Hence, if $x = 1$ then $y = 1/2$. This means that one point that is on the line is identified by the address of $(1, 1/2)$. Now we select a second number for x, say $x = -3$, and we do the process all over again.

$$\begin{aligned} 4(-3) - 2y &= 3 && \text{substituting } -3 \text{ for } x \\ -12 - 2y &= 3 && \text{multiplying 4 times } -3 \\ -2y &= 15 && \text{moving } -12 \text{ to the right and adding} \\ y &= -15/2 && \text{dividing by } -2 \text{ and simplifying.} \end{aligned}$$

Therefore, a second point on our line is $(-3, -15/2)$. Since any two points determine a straight line, we can plot these two points and then draw the straight line that passes through them as in Figure 38.

But straight lines are only straight lines, and they are going to get boring if we don't move on. Since we know that linear equations always produce lines it is natural to ask what the next higher level of equations produces. The next higher level equation is called a quadratic equation because, at most, each term has the unknowns raised to the second power. In other words, our terms might include expressions involving x^2, y^2, or xy. Such terms are called quadratic because they involve the power of two, or they have two linear terms multiplied together. Such equations might also involve a combination of linear terms, such as x or y, and a constant term. The most general expression for a quadratic equation in two unknowns is:

$$Ax^2 + Bxy + Cy^2 + Dx + Ey + F = 0$$

where all the capital letters, A through F, represent coefficients. If A, B, and C are all zero then the equation becomes simply a linear equation. However, if one or more of these terms is not zero, then we have a true quadratic equation.

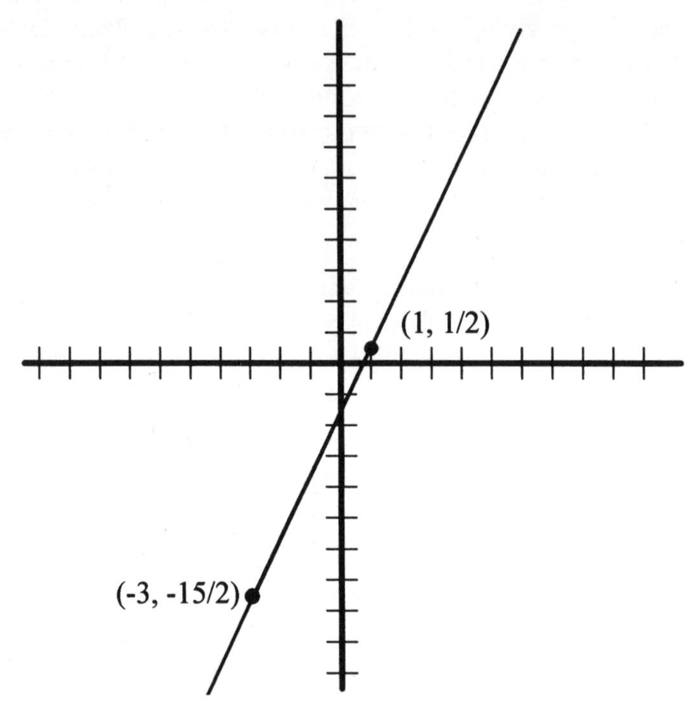

FIGURE 38. The linear equation $4x - 2y = 3$ graphed onto the Cartesian coordinate system.

If we plot such quadratic equations onto our Cartesian coordinate system, what kinds of pictures do we get? Hold on to your seat, for this is absolutely amazing. Every quadratic equation generates a conic section (parabola, ellipse, or hyperbola), and every conic section has a corresponding quadratic equation. Some of the quadratic equations produce what we call degenerate curves. For example, if we take our two cones and pass a plane through the exact point where their tips touch, then we don't get a regular conic curve, but just a single point. It is also possible to pass a plane through two cones in such a manner that the plane cuts out only a single straight line. This is another degenerate conic section.

Therefore, the very conic sections that Apollonius studied in ancient Greece turn out to be the curves generated by quadratic equations. Thus it is exceedingly easy to study conic sections because we can manipulate equations according to algebraic rules, and the results will still be valid when we graph them back onto a Cartesian coordinate system.

To demonstrate the power of Fermat's method to represent curves, consider a circle whose center is located at the origin of a Cartesian coordinate plane. We know that the definition of a circle is the loci of points equidistant from a single point—its center. We can use this definition to generate the equation for the circle. Notice in Figure 39 that we have drawn such a circle and we have drawn a radius, R, from the center to the circle itself. Now we are looking for the equation in x and y such that every point on the plane that is on that circle is also a solution to the equation, but no other points are solutions. We can use the Pythagorean theorem to represent the distance R in terms of x and y, for R is the hypotenuse of a right triangle, while x and y are the two legs. Thus we have $x^2 + y^2 = R^2$. For example, if we have a circle whose center is at the origin and has a radius of 2, the corresponding quadratic equation becomes $x^2 + y^2 = 4$. The

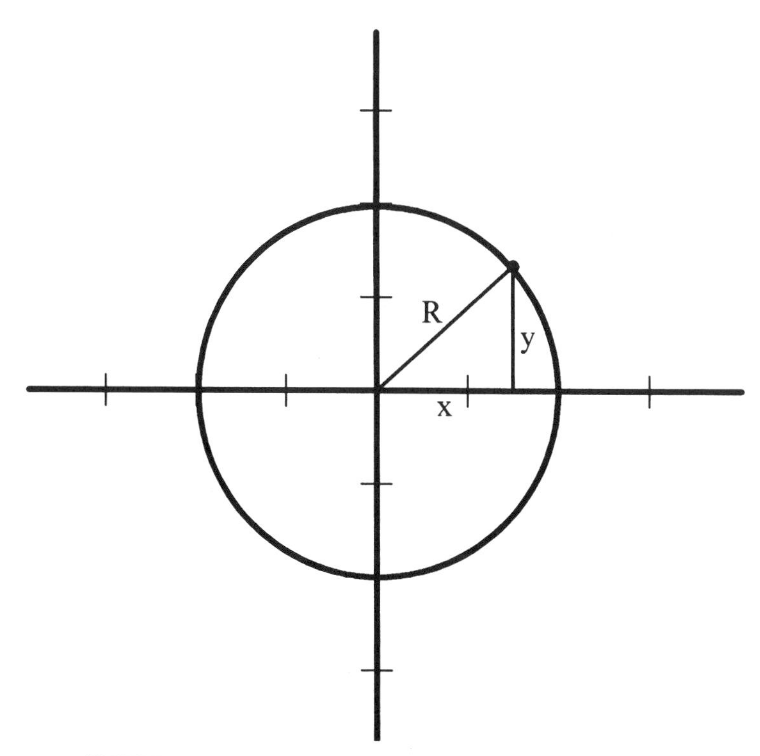

FIGURE 39. Generating the equation of the circle $x^2 + y^2 = R^2$.

various points of the circle, expressed as (x, y), satisfy this equation and only such points satisfy it.

But what of a circle that is not located with its center at the origin? And what about the parabola, ellipse, and hyperbola? We obviously need to examine the various forms of quadratic equations more closely.

The simplest quadratic equation that produces a parabola is $y = x^2$. This parabola has been drawn in Figure 40. First we want to point out several interesting features possessed by every parabola. The lowest point on the parabola in Figure 40 is called the vertex of the parabola. If we had a parabola which opened down instead of up, the vertex would be its highest point. Also, for every parabola there exists a line which passes through the vertex and divides the parabola into two symmetric halves. This line is called the axis of symmetry, and in Figure 40 it happens to be the vertical axis.

In addition, every parabola has a special point inside the parabola's bowl on the axis of symmetry. If we shine light straight into a parabola, the light reflects off of the curve in such a manner that it concentrates at this point. If we locate a light source at this point, then its light will reflect off the bowl of the parabola and out in a straight beam. This point is called the

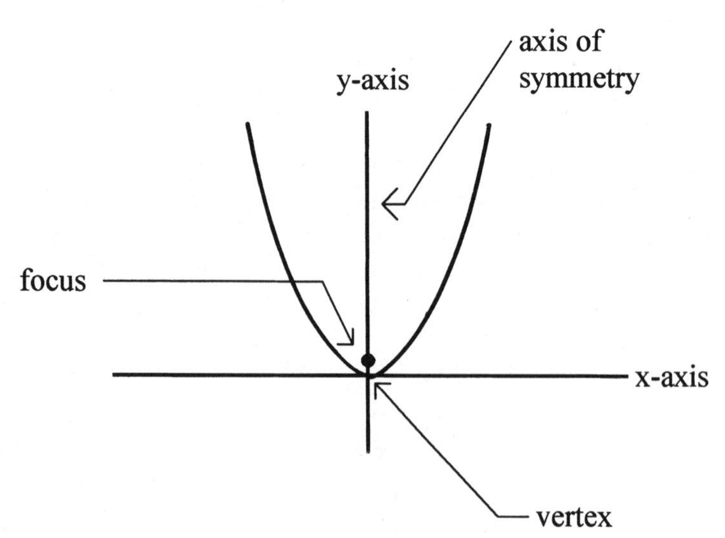

FIGURE 40. The parabola $y = x^2$. The y-axis is the axis of symmetry, the vertex is at the origin, and the focus is ¼ unit above the vertex on the y-axis.

focus and is what makes the parabola so useful in the modern world. We all own our personal parabolas: the headlights of our cars, the reflectors on our flashlights, the satellite TV dishes we use. All these are generated by taking a parabolic curve and rotating it to make a dish shape, called a paraboloid of revolution.

We can start with our parabola in Figure 40, represented by the quadratic equation $y = x^2$ and make several changes that move the parabola around on the plane. These changes are:

1. Orientation: We can make the parabola in Figure 40 open down instead of up by just changing the sign in front of the x term. For example, the parabola $y = -x^2$ is plotted in Figure 41a. If we want a parabola that opens to the left or right, we just interchange the x and y terms. For example to get a parabola to open toward the right we use the equation $x = y^2$.

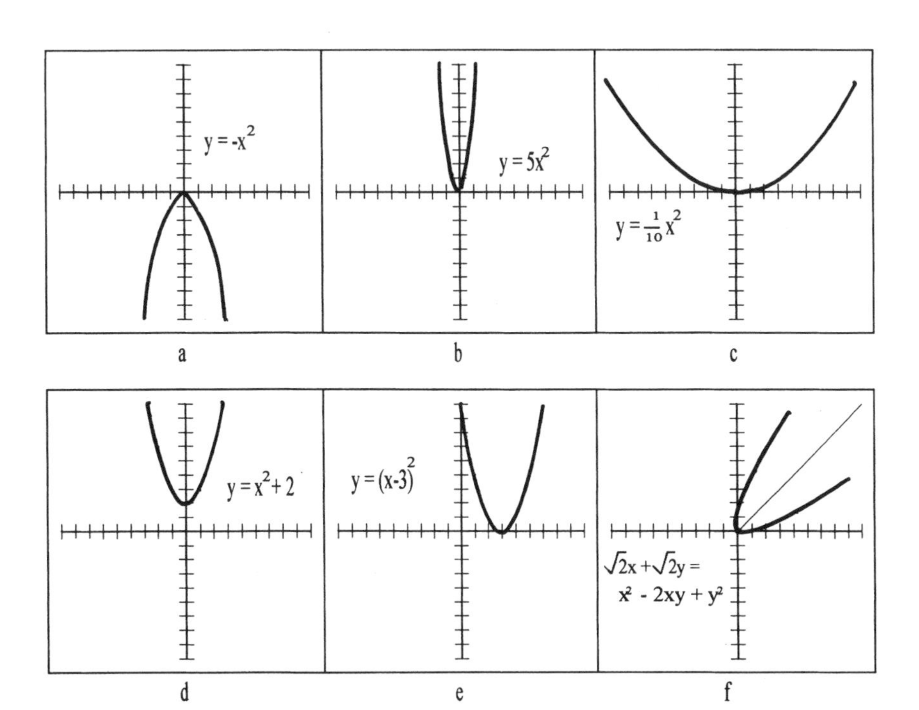

FIGURE 41. The graph of the parabola $y = x^2$ is moved around the Cartesian plane through specific adjustments.

2. Shape: We can change the shape of the parabola by changing the size of the coefficient in front of the x term. If we change the x coefficient from 1 to a larger number we increase the steepness of the curve, causing the curve to fold up toward the axis of symmetry. But if we make the x coefficient smaller than 1, then we cause the curve to gently lie down closer to the x axis. In Figure 41b we have plotted the parabola $y = 5x^2$, and in Figure 41c we have $y = (1/10)x^2$. Notice how the first parabola folds up closer to the y axis while the second opens much slower.

3. Shift up or down: We can move the parabola up and down by simply adding a constant to or subtracting a constant from the right side of the equation. For example, we move the parabola $y = x^2$ up two units by writing it as $y = x^2 + 2$ (Figure 41d). In like fashion, to move the parabola down two units, we simply subtract two from the right to get $y = x^2 - 2$.

4. Shift right or left: We can move the parabola either toward the right or the left by replacing the x term with $x \pm c$, where c is the amount we wish it to move. For example, if we want to move the parabola $y = x^2$ toward the right three units, we replace x with $x - 3$ to get, $y = (x - 3)^2$ (Figure 41e). To move it three units toward the left, we replace x with $x + 3$.

5. Rotate: The most difficult move is to rotate the parabola. In the parabola $y = x^2$ the axis of symmetry is a vertical line. If we want this line to be other than vertical or horizontal we must include a term into its equation containing xy. Hence, to rotate $y = x^2$ by 45 degrees we need to change the equation by adding several additional terms in x and y, or $\sqrt{2}x + \sqrt{2}y = x^2 - 2xy + y^2$ (Figure 41f). However, to determine the proper terms to add to a quadratic equation to cause a rotation is not simple, so we will leave until later just how this is done.

The five procedures above allow us to move and change a parabola into any form we want. Based on the ideas we have presented, we can take the general equation of a parabola which has not been rotated (lacks an xy term) and rewrite it in a special form which will tell us all about the parabola involved. It is much like doing a crossword puzzle—great fun! The special form for the parabola when the xy term is missing is:

$$4p(y - k) = (x - h)^2$$

Now you may be asking, what the heck is this h, k, and p? The h and k represent the coordinates of the vertex of the parabola. Hence, (h, k) is the exact point in the plane that represents either the lowest or highest point on the parabola (Figure 42). The number represented by p is the distance of

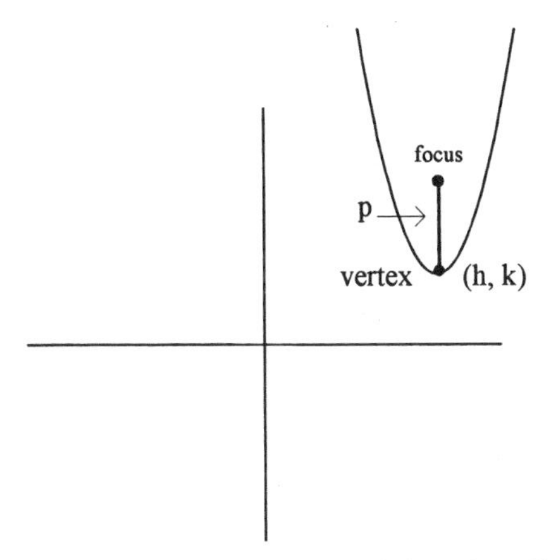

FIGURE 42. The general form of the parabola, $4p(y - k) = (x - h)^2$.

the focus from the vertex. Therefore, if we can find the values of h, k, and p we immediately know where this parabola is located, where the focus is, and based on the sign of p, whether it opens up or down. For example, let's consider the parabola which has the equation $y = (1/2)x^2 - x + 3.5$. Using a method from algebra called "completing the square" we can rewrite this equation in its special form of $2(y - 3) = (x - 1)^2$. An example of completing the square can be found in the Endnotes.[9]

From our rewritten equation we know at once that the vertex of this parabola is located at (h, k), which in this case is (1, 3). Since $4p = 2$ we can see that $p = ½$. Since p is positive, we know that our parabola opens up and that the focus is located on the axis of symmetry ½ unit above the vertex at (1, 3.5). This parabola is plotted in Figure 43.

We can also accomplish the five procedures with both ellipses and hyperbolas; that is, we can change their shapes, change their orientations, shift them up or down, right or left, and we can rotate them. Do the ellipse and hyperbola have lines that play the same role as the axis of symmetry in the parabola? Yes, look at Figures 44 and 45, where we have drawn in these two curves and labeled the interesting characteristics. With an ellipse we have two lines of symmetry: the major axis and the minor axis. Both axes pass through the center of the ellipse, with the major axis the longer of the two. The ellipse also has two foci located symmetrically on the major axis on either side of the center, and two vertices where the major axis inter-

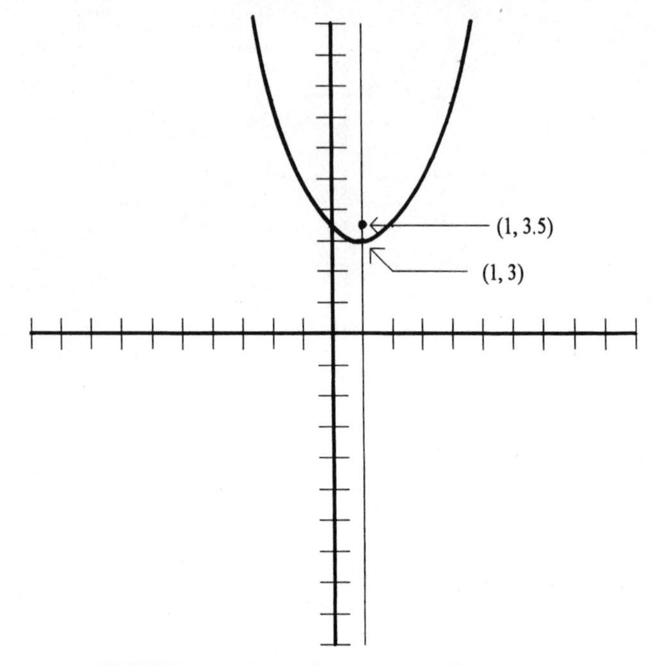

FIGURE 43. The parabola $y = (1/2)x^2 - x + 3.5$.

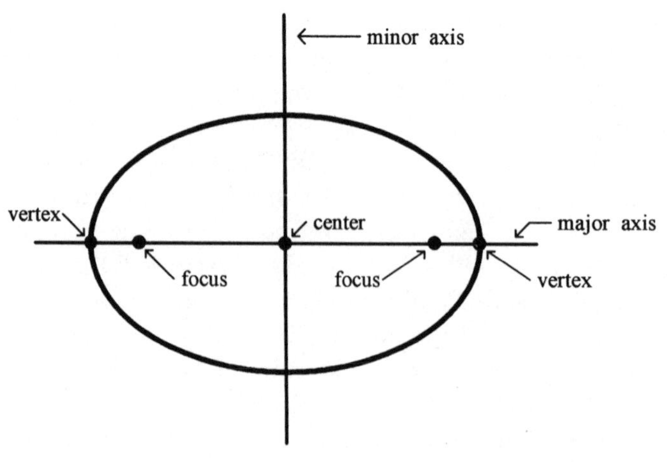

FIGURE 44. The ellipse, with its center, major and minor axes, two vertices, and two foci.

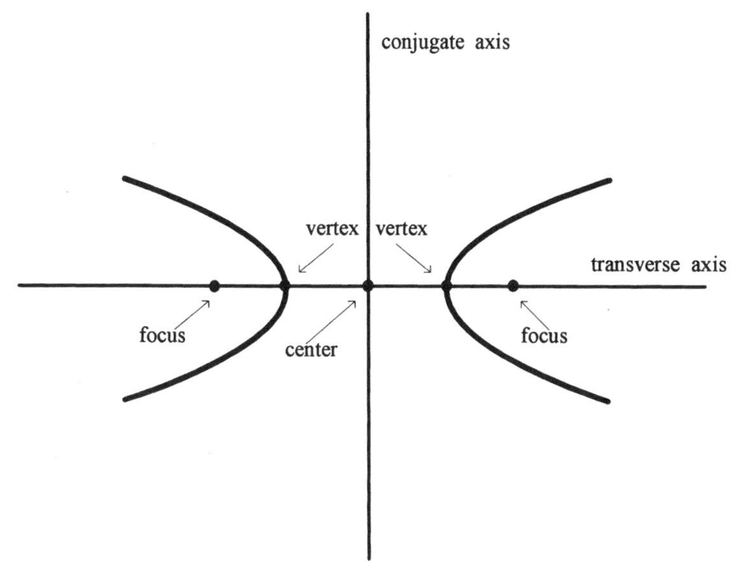

FIGURE 45. The hyperbola, with its center, transverse and conjugate axes, two vertices, and two foci.

sects the ellipse. The hyperbola also has two axes of symmetry: the transverse axis and the conjugate axis. Located on the transverse axis are two foci, two vertices, and the center.

We can get a better grasp of both the ellipse and hyperbola if we begin as we did with the parabola and consider curves that have not been rotated and have their centers located at the origin of a Cartesian plane. An ellipse whose center is at the origin is generated by the equation:

$$\frac{x^2}{a^2} + \frac{y^2}{b^2} = 1$$

Now you're possibly wondering what the a and b terms are. They define the numbers that control the shape of the ellipse. The a term is the distance from the center of the ellipse to its two vertices along the major axis. The b term is the distance from the center of the ellipse to where the minor axis intersects the ellipse. If a is bigger than b, then the ellipse has a horizontal major axis, and the ellipse lays on its side. If b is larger than a, then the major axis is vertical, and the ellipse rests on end. The greater the difference between a and b, the more elongated the ellipse. When a equals b, then the above equation becomes that of a circle. Hence, a circle is just an ellipse whose major and minor axes are the same size. Of course we know that the orbit of all the planets, including the Earth, are ellipses. For each orbit, the Sun is not located at the center of the ellipse, but at one of the foci on the major axis. The Earth's orbit is almost circular, with the distance between the two foci approximately 3 million miles, which is the difference between Earth's closest approach to the sun and its farthest approach.

We have determined how to change the orientation of the ellipse, since we can just interchange the a and b in the equation. We can also change the shape of the ellipse by changing the relative sizes of a and b. How do we move the ellipse around? We do it the same way as with the parabola, and when we're done we end up with the general equation of an ellipse which has not been rotated (no xy term).

$$\frac{(x-h)^2}{a^2}+\frac{(y-k)^2}{b^2}=1$$

As with the parabola, the h and k identify where the center of the ellipse is. The two major vertices are a distance equal to a from the center along the major axis, while the two minor vertices are a distance equal to b from the center. All we need now is to know where to place the two foci. This is easy because the distance from the center to each foci is related to the a and b terms by the following formula: $f^2 = a^2 - b^2$. Therefore, from the above special form of the elliptic equation, we can determine just where our ellipse is and what it looks like.

The hyperbola is identical to the ellipse, except that the two terms on the left have opposite signs, instead of both being positive. Therefore, the general equation of an nonrotated hyperbola is:

$$\frac{(x-h)^2}{a^2}-\frac{(y-k)^2}{b^2}=1$$

The hyperbola has two separate branches. The orientation of the branches, whether they open up and down, or right and left, is controlled by which of the terms on the left is positive and which is negative. If the term containing y is negative, then the two branches open right and left as in Figure 45. However, if the term containing the x is the negative term, then the two branches open up and down.

The hyperbola offers something extra special. The a and b terms again

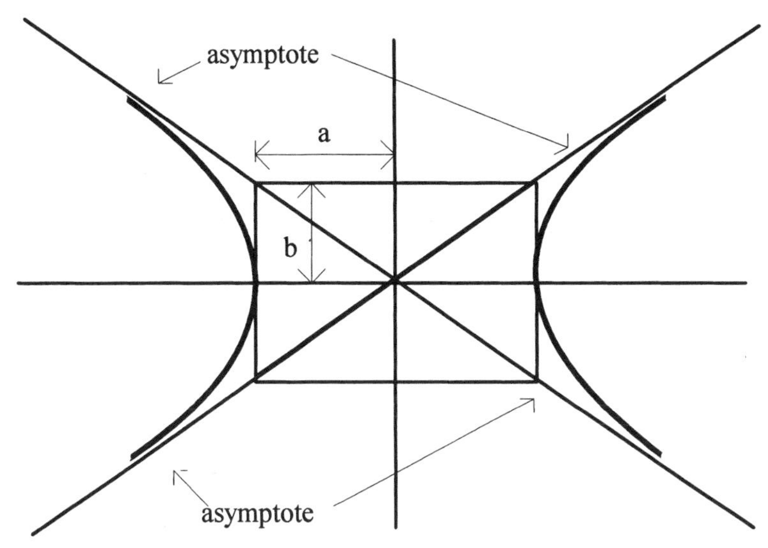

FIGURE 46. A hyperbola showing its asymptotic lines. If the hyperbola's equation is $x^2/a^2 - y^2/b^2 = 1$, then the asymptotic lines are the diagonals of the box whose length is $2a$ and whose height is $2b$.

control the shape of the hyperbola. We can use a and b to construct a box between the two branches of the hyperbola as in Figure 46. Now we draw two straight lines, each passing through diagonal corners of this box. These lines, called asymptotic lines, define the shape of the two hyperbolic branches. These are lines dearly loved by mathematicians because of their unique qualities. In our graph of a hyperbola, the curve will get ever closer to the asymptotic line, but will never touch it. We can understand this characteristic if we look at the simple equation $y = 1/x$, which has been plotted in Figure 47. Notice that as we move to the right where the values on the x-axis grow ever larger, the value of y grows smaller. Y will never become zero because we will always have a fraction with 1 in the numerator and some large number in the denominator. However, we can get the curve defined by $y = 1/x$ as close to the x-axis as we wish. As we move toward the right, the curve is getting ever closer to the x-axis but will never touch it. The x-axis is therefore an asymptotic line to the curve. In just the same way, our diagonal lines in Figure 46 are asymptotic to the hyperbola.

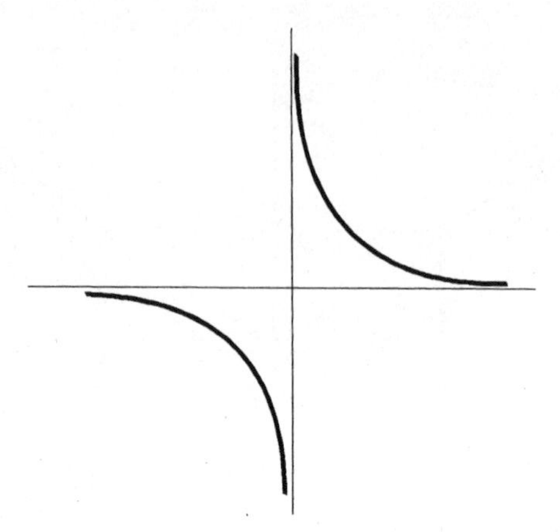

FIGURE 47. The hyperbola $y = 1/x$ with the x-axis and y-axis as asymptotic lines.

This brings us to an interesting side note. The equation $y = 1/x$ can be rewritten as $xy = 1$. But wait! This must be some kind of conic section for it is a quadratic equation that contains at least one quadratic term and a constant. What conic curve is it? Look again at Figure 47. This is just the graph of a hyperbola that has been rotated 45 degrees. If we rotated it back to eliminate the xy term we would get the equation:

$$\frac{x^2}{2} - \frac{y^2}{2} = 1$$

As with the parabola, the ellipse and hyperbola have interesting reflective properties, too. Waves originating at one focus of an ellipse reflect off the ellipse to concentrate at the other focus. For example, in a swimming pool in the shape of an ellipse we can drop a stone at one focus, and the waves will bounce off the pool's sides to concentrate at the second focus (Figure 48). This characteristic of ellipses is used in modern medicine to break up kidney stones.[10] The patient is positioned in an elliptic pool, called a lithotripter, so that his kidney stone is located at one focus while high energy sound waves are produced at the other focus. These waves bounce off the lithotripter's walls to refocus at the focal point in the patient and break up the stone.

A domed room made in the shape of an ellipse rotated through a half plane is called a whispering gallery and reflects sound from one focus to

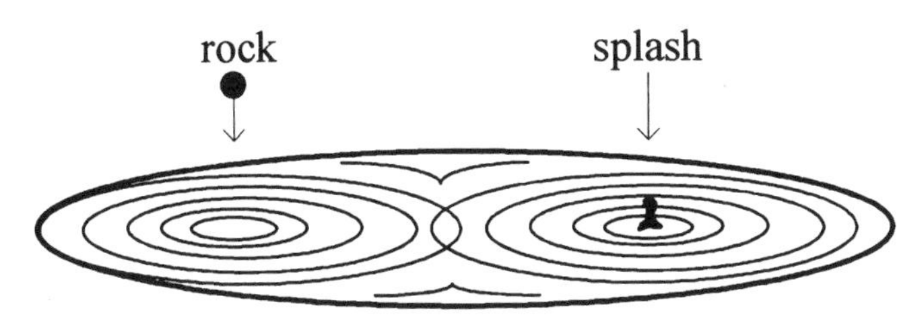

FIGURE 48. A pool in the shape of an ellipse reflects waves from one focus to the other focus. This principle is used in modern medicine to break up kidney stones.

the other (Figure 49). The Mormon Tabernacle building, home of the Mormon Tabernacle Choir, on the temple grounds in Salt Lake City, Utah is just such a room, as is the Statuary Hall in the Capitol building in Washington, D.C. A whisper at one focus of such rooms can be plainly heard at the other focus.

The reflective property of the hyperbola is very strange. A beam of light aimed directly toward the focus from the convex side of a hyperbolic mirror (Figure 50) will reflect off the mirror and back toward the other focus. The reflective property of the hyperbola is used in conjunction with that of the parabola in a Cassegrain telescope. Invented in 1672 by Sieur Cassegrain, this design is incorporated in the Hubble Space Telescope (Figure 51).

You have probably learned more about parabolas, ellipses, and hyperbolas than you ever knew you wanted to know. However, the importance

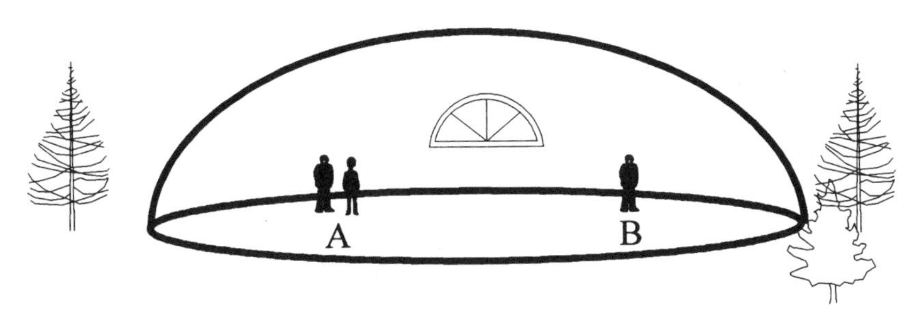

FIGURE 49. A whispering gallery is a domed room made in the shape of an ellipse rotated through a half plane. Whispers from one focus (A) reflect off the walls to gather at the other focus (B).

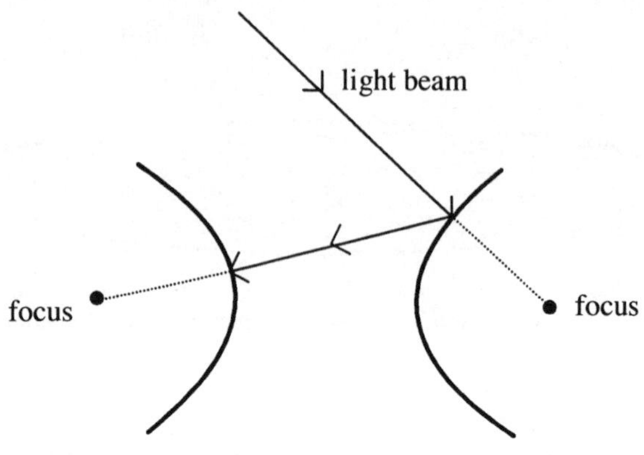

FIGURE 50. The reflective property of the hyperbola. Light directed toward the focus from the convex side of a hyperbolic mirror reflects back toward the other focus.

of these curves can't be overstated. Physics demands an understanding of motion, and these curves describe the three basic kinds of curvilinear motion in our universe. How could we plot the location of the Earth or the planets, or put satellites into orbit, or even receive a T.V. signal without conic sections? Yet, these curves are beautiful in their own right. They have a delicate grace which catches and pleases the eye.

So far we have limited ourselves to two dimensions. However, we can extend the idea of the conic sections to three dimensions by looking at the curves generated by quadratic equations with three unknowns (x, y, and z)

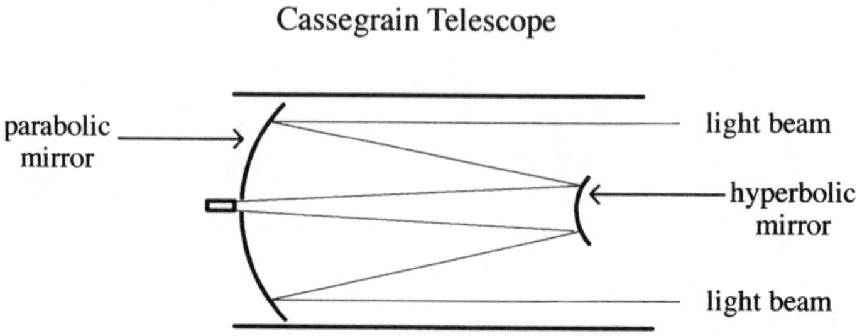

Cassegrain Telescope

FIGURE 51. The Cassegrain telescope, invented in 1672 by Sieur Cassegrain, uses a secondary mirror that is hyperbolic in shape. This arrangement is used in the Hubble Space Telescope.

where each unknown represents a different axis. The general equation for conic solids whose axes have not be rotated is:

$$Ax^2 + By^2 + Cz^2 + Dx + Ey + Fz + G = 0$$

This is analogous to our two-dimensional equation with the z terms added. First we consider the conic solid called the Ellipsoid:

$$\frac{x^2}{a^2} + \frac{y^2}{b^2} + \frac{z^2}{c^2} = 1$$

The Ellipsoid has the beautiful shape shown in Figure 52.

Next is the Hyperboloid of One Sheet:

$$\frac{x^2}{a^2} + \frac{y^2}{b^2} - \frac{z^2}{c^2} = 1$$

Here we see that the only difference between the Ellipsoid and the Hyperboloid of One Sheet is that the positive sign has become negative in front of

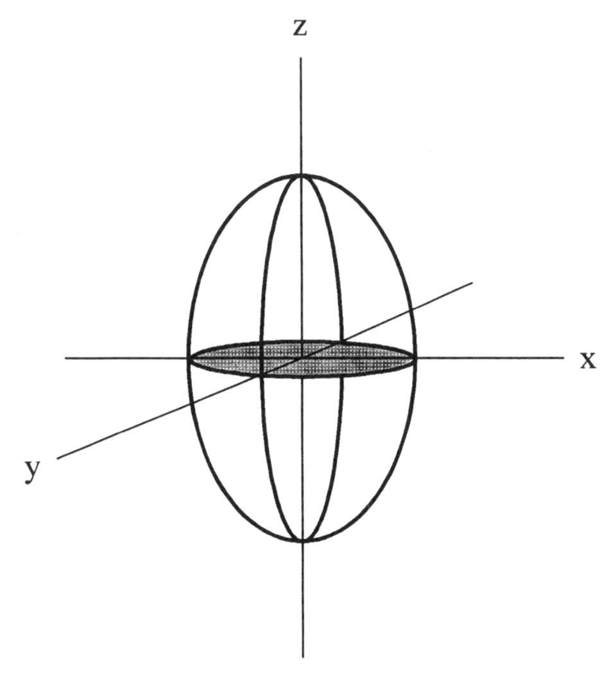

FIGURE 52. The Ellipsoid, $x^2/a^2 + y^2/b^2 + z^2/c^2 = 1$.

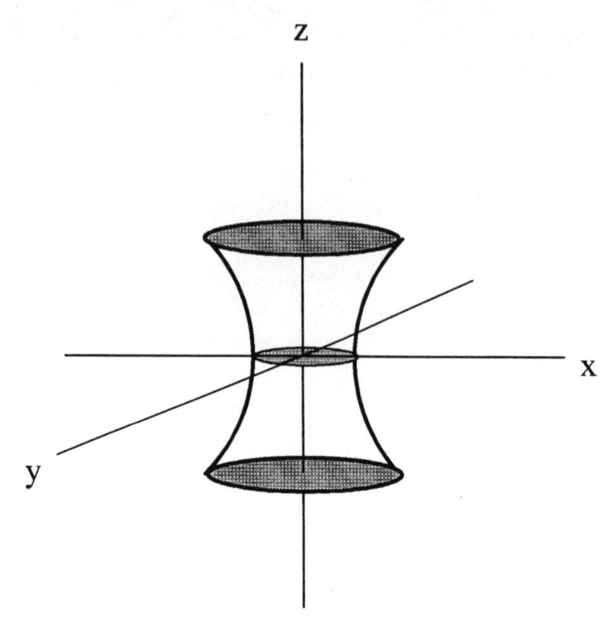

FIGURE 53. The Hyperboloid of One Sheet, $x^2/a^2 + y^2/b^2 - z^2/c^2 = 1$.

the z term. This conic is shown in Figure 53. The two-dimensional curve generated when a solid intersects a plane is called a trace. The trace generated by this Hyperboloid and the xy-plane is a circle if $a = b$, but is an ellipse if $a \neq b$. The trace made in both the xz-plane and the yz-plane is a hyperbola.

Next we have a Hyperboloid of Two Sheets (Figure 54), with the following equation:

$$-\frac{x^2}{a^2} - \frac{y^2}{b^2} + \frac{z^2}{c^2} = 1$$

The difference between this Hyperboloid and the previous one is that now we have two negative terms instead of just one. One Sheet means the Hyperboloid is a single three-dimensional figure. Two Sheets means the Hyperboloid is divided into two symmetrical figures separated by some finite distance. The Hyperboloid of Two Sheets yields traces parallel to the xy-plane that are ellipses, while the traces in both the xz-, and yz-planes are hyperbolas.

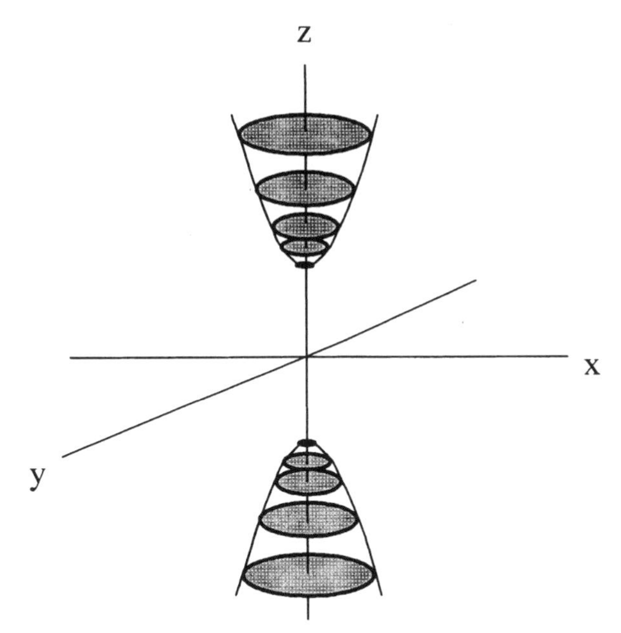

FIGURE 54. The Hyperboloid of Two Sheets, $-x^2/a^2 - y^2/b^2 + z^2/c^2 = 1$.

Very similar to the Hyperboloid of Two Sheets is the Elliptic Paraboloid (Figure 55), which has a single cup and the following equation:

$$\frac{x^2}{a^2} + \frac{y^2}{b^2} = \frac{z}{c}$$

With this solid, we have the z term as linear instead of quadratic. If $a = b$ in the above equation then we get a Paraboloid of Revolution.

Next we have a Hyperbolic Paraboloid (Figure 56) with the following equation.

$$\frac{x^2}{a^2} - \frac{y^2}{b^2} = \frac{z}{c}$$

This is identical to the previous equation except the y term is negative. The traces in the xy-plane are just intersecting lines. Traces generated by planes parallel to the xy-plane are hyperbolas, while the xz- and yz-plane traces are parabolas.

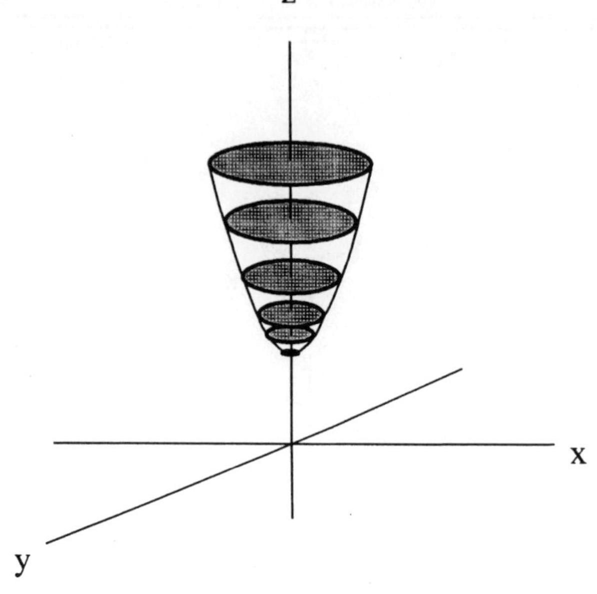

FIGURE 55. The Elliptic Paraboloid, $x^2/a^2 + y^2/b^2 = z/c$.

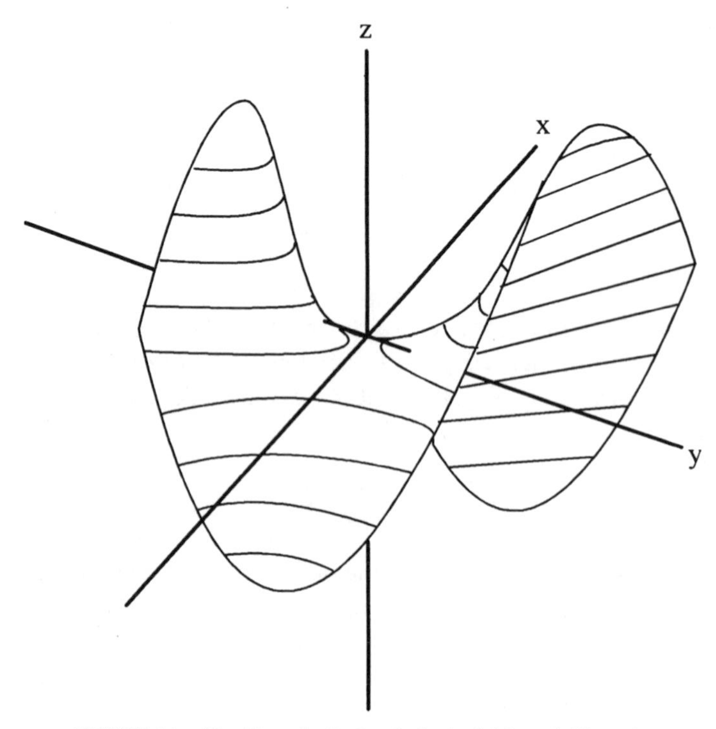

FIGURE 56. The Hyperbolic Paraboloid, $x^2/a^2 - y^2/b^2 = z/c$.

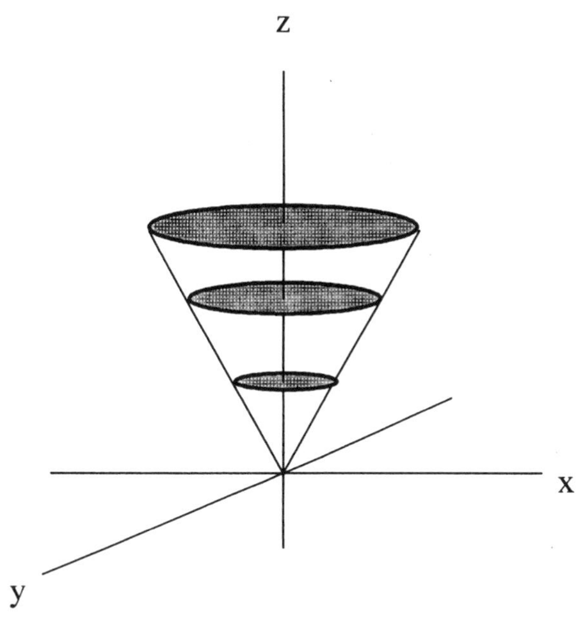

FIGURE 57. The Cone, $x^2/a^2 + y^2/b^2 - z^2/c^2 = 0$. This equation is identical to the Hyperboloid of One Sheet except the constant term is 0 rather than 1.

We end our journey into three dimensions with the very solid that excited Apollonius in the first place, the Cone (Figure 57).

$$\frac{x^2}{a^2} + \frac{y^2}{b^2} - \frac{z^2}{c^2} = 0$$

FERMAT AND DESCARTES: A FINAL NOTE

While it has been customary to give primary credit to René Descartes for the origins of analytical geometry, we would be amiss not to give equal credit to Fermat. While Descartes improved symbolic algebra, it was Fermat who fully utilized perpendicular axes and worked with the graphs of so many different curves. However, neither of their contributions to humankind should be restricted to their analytic geometry, for they both contributed heavily in other fields. Fermat, especially, contributed to mathematics. Both he and Descartes were instrumental, along with numerous other mathematicians, in developing the problems of calculus before Newton and Leibniz finally discovered the fundamental theorem of calculus. Fermat also corresponded with Blaise Pascal (1623–1662) to originate the theory of probability. Moreover, he is credited with independently establishing modern number theory. Though Descartes may have been the better philosopher, Fermat was the better mathematician.

5

Opening the Door

We are now going to explore two closely connected subjects: infinite series and limits. The field of infinite series is unbelievably interesting by itself, and the concept of a limit is critical to understanding calculus. Explaining the ideas behind infinite series and limits is somewhat difficult, not because the ideas are difficult, for they are both simple and beautiful. The difficulty arises from my nervous anticipation. Explaining such profound secrets to others is to be ushered into the Temple of Truth where I must address the Goddess of Rational Knowledge. To be in her presence, and to speak, gives me pause. However, the rewards are worth any initial discomfort, for soon we will be in a magical land that will bestow great pleasure and power upon the faithful.

SEQUENCES

A number sequence is simply an ordered list of numbers. The study of sequences and series is very old, with references in both the earliest Egyptian and Babylonian mathematical works. We are intimately familiar with number sequences because at a very young age we all learned how to count by memorizing the counting numbers from one to ten. Once we have this mastered we learn that we can make longer number sequences by counting even higher. Therefore, the best example of a number sequence is the counting number sequence: 1, 2, 3, ... The individual numbers (terms) are not added or subtracted, just listed. Recognizing that the counting numbers form a sequence demonstrates that sequences can be infinitely long. How do we know it is infinitely long? Children prove it to themselves, possibly in some early childhood argument with a friend or sibling.

"Numbers go on forever," claims the ten-year-old child.

"No they don't," retorts the seven-year-old.

"If they don't go on forever, where do they stop?"

The little one thinks for a moment. "They stop at a billion zillion—that's the last number."

The older child, looking victorious, responds: "Oh yeah? What if I add one to your billion zillion? What do I get then? I get a billion zillion plus one. So, I win—numbers go on forever."

Now we all have to stop and think. Can numbers actually go on forever? If they ever did stop, you could force them to continue by just adding more numbers. Even if we have forgotten this as adults, we have all certainly experienced the same kind of argument; and finally, because of the pure logic of the idea, we acquiesce and accept the infinity of numbers.

It is this very property of sequences (and their cousins, the series) that makes them interesting. If we could not comprehend the infinity of our counting sequence, if every set of numbers had to be finite, then modern mathematics would not exist. I find it astounding that humans can conceive of the infinite at all. Nothing in our surroundings suggests an infinite. We see only to the horizon. Our family and friends are limited. All of our possessions are limited. Our time here on earth is finite. How, then, did we ever come up with such a profound idea as the infinite? As already pointed out, the idea of infinity is not new, but stretches back to at least the legends of Gilgamesh, 2000 B.C.

Others, too, have wondered how limited human beings can conceive of the infinite. René Descartes (1596–1650), the cofounder of analytical geometry, used this idea for a proof of the existence of God: humans, being limited beings, can generate only limited ideas. Because we possess the concept of infinity, some infinite being must have given us that idea—that infinite being is God. Of course, this argument depends on the idea that a limited being can generate only a limited idea.

We are not interested here in the infinity which may be found in nature, either of extension (space) or time. Rather, we are interested in the infinity within the world of ideas. Just how ideas partake of existence has been a puzzlement to philosophers since the time of Plato. Ideas do not have a material reality such as rocks and stars, nor are they restricted to a particular space and time. Yet, they do share in some fashion in existence or we would not be able to discuss their attributes.

Our infinite number sequence, as an idea, does not depend upon any physical infinity, but exists as long as its attributes do not contain a contradiction. Some mathematicians have claimed that to allow for infinite mathematical objects does introduce contradictions into mathematics, and they attempt to do mathematics without using this concept. So the battle between those who love the infinite and those who shy from it continues. For what follows, we are going to gladly embrace the infinite.

If the individual terms of a sequence continue to grow in size, with no limit to their eventual magnitude, we say the sequence diverges. For example, the number sequence itself diverges since its terms continue to grow ever larger. If the individual terms never grow in size beyond some fixed number, we say the sequence is convergent or that it converges to that fixed number. For example, the sequence 1, ½, ¼, ⅛, . . . has terms that are actually decreasing in size, hence it is a converging sequence. In like fashion, the sequence 1, 1, 1, 1, . . . is neither increasing nor decreasing, and is convergent.

One of the most famous sequences is called the *Fibonacci sequence*: 1, 1, 2, 3, 5, 8, 13, . . . where each new number is just the sum of the two previous numbers. This sequence was suggested (and named after) Leonardo of Pisa (1170–1240), who was also known as Fibonacci. He wanted to know how many pairs of rabbits will be produced each year if we begin with a single pair which mature during the first month and then produce another pair of rabbits every month after that. The Fibonacci sequence of numbers answers that question. Since the Fibonacci sequence is related to how large certain populations grow, it is frequently found in nature. So much interest has been generated regarding this sequence that a Fibonacci Society has been founded to study and record its many surprising properties in the *Fibonacci Quarterly*.[1]

Sequences in general have produced so much interest among mathematicians that a sequence registry for integer sequences now exists on the Internet.[2] It is called *The Encyclopedia of Integer Sequences*, and is maintained by Neil J. A. Sloane, who is a technology consultant for AT&T Labs-Research. This registry contains over 25,000 different number sequences. You can visit this Internet page and enter a number sequence which you are curious about. The page's search engine will locate and identify similar sequences for you. You can also submit your own sequence for listing. Sloane and S. Plouffe have published the 5,487 most interesting sequences in book form.[3] *The Encyclopedia of Integer Sequences* lists the Fibonacci sequence as sequence number A000045.

SERIES

The series is closely related to the sequence, for a series is just the sum of a set of numbers. Therefore, whenever we have a number sequence, we can turn it into a series by just adding its individual terms. While 3, 4, 7 is a sequence with three terms, the sum $3 + 4 + 7$ is the corresponding series with three terms. If infinite sequences exist we can use them to build infinite series. One such infinite series would be the sum of all the counting numbers, or $1 + 2 + 3 + 4 + \ldots$, using three dots, an ellipsis, after the 4 to indicate that the addition of numbers continues without end. We cannot actually add together all counting numbers nor can we see their sum written out, for that would require an infinitely long page filled with numbers extending forever into space.

Does our infinite series, consisting of all counting numbers, add up to some specific number? It can't because any individual number is a term in the series. Hence, the series is larger than any individual number, if the word "larger" has any meaning in this situation. Therefore, whatever such a series adds to, the sum is not another counting number. When an infinite series is larger than every counting number (or every real number) then we use the same word we used with sequences and say the series is divergent. We can now see that there must be infinitely many divergent infinite series, because we can construct an infinite number of them by various arrangements of the number sequence. Any infinite series whose individual terms are all whole numbers will add together to be larger than any finite number—it will diverge. We can even make a divergent infinite series from an infinite number of ones: $1 + 1 + 1 + \ldots$.

We need a better notation for an infinite series than just writing down the first few terms. The simplest notation is to use an *S*. We can place a number as a subscript to the *S* to indicate how many numbers are in the series, and we can use the symbol for infinity if it is an infinite series. Hence, we have:

$$S_5 = 3 + 5 + 7 + 9 + 11$$

$$S_\infty = 1 + 4 + 9 + \ldots$$

In this example we have designated S_5, which contains five odd numbers in sequence beginning with 3. S_∞ contains all the squares of the positive integers—an infinite set of numbers. We can also use a subscript if we just want to distinguish between two series, such as the series S_a and S_b.

However, it would be useful to have another symbol for a number series, one containing more information than the series' size, or just its first few terms. To do this we use the Greek letter, capital sigma (Σ), standing for a sum of terms. Each term in the series is associated with an index number that identifies where that term falls within the series. We define the beginning index number below the sigma, and the ending index on top. If the series is infinite, we place the symbol for infinity (∞) above the sigma. The short notation for our infinite series of counting numbers is:

$$\sum_{n=1}^{\infty} n = 1 + 2 + 3 + 4 + \ldots$$

That is, we begin with an index, n, equal to 1 and then we continue to generate each new term by adding 1 to n. The infinity symbol above the sigma tells us there is no end to this process. This new notation has several advantages. Not only is it shorter to write down, but by using it we also make the series well defined. Consider the two series which have been defined only by their first three terms:

$$S_a = 2 + 4 + 8 + \ldots$$
$$S_b = 2 + 4 + 8 + \ldots$$

Now the two series, S_a and S_b, appear to be the same since their first three terms are identical. However, watch how I can extend them:

$$S_a = 2 + 4 + 8 + 16 + 32 + 64 + \ldots$$
$$S_b = 2 + 4 + 8 + 14 + 22 + 32 + \ldots$$

Now we see that the two series are very different. The first series is just the increasing powers of 2, while the second series begins with 2 and then generates each new number by adding consecutive even numbers. Hence, the first difference is 2, the second (between 4 and 8) is 4, the third is 6, etc. This example illustrates that just listing a finite number of terms is insufficient to delineate between two different series. What we actually need is a definition of how each successive term is to be generated. We do this using the sigma notation. After the sigma we write the formula we use to generate each successive term in the series. For the first series above we have:

$$S_a = \sum_{n=1}^{\infty} 2^n = 2 + 4 + 8 + \ldots$$

Following the sigma above is the formula for generating the series we desire, 2^n. If we substitute the number 1 for n in 2^n we get $2^1 = 2$, which is the first number in the series $2 + 4 + 8 + \ldots$. When we substitute the number 2 into 2^n we get $2^2 = 4$, or the second term in the series. Hence, when we want the kth term in the series, we substitute k for n in 2^n and find our term. Thus we see that $\Sigma\, 2^n$ generates all the terms of the desired series by consecutively substituting in the index numbers from the number series: 1, 2, 3, ...

For the second series we have the following formula:

$$S_b = \sum_{n=0}^{\infty} (n^2 + n + 2) = 2 + 4 + 8 + \ldots$$

Again, we simply substitute consecutive index numbers for n from the number series to get the successive terms of the series we are interested in. In this case we begin not with 1 but with 0, as is indicated by the equation n=0 found under the sigma.

Now both series are completely defined, for we know how to generate every term within these two series. However, for aesthetic reasons we will frequently list both the sigma notation and the first few terms. When we see successive terms they often reveal a symmetry and progression that is very pleasing.

We know that divergent series exist, those series whose sums are larger than any counting number. Are there series whose sum is a finite amount? Consider the problem of walking one mile to the corner store by walking in ever shorter intervals. First, we walk a half mile and then stop. For the next interval we walk half of the remaining distance, which is a quarter mile, and stop again. We continue to do this, always walking half the remaining distance and then stopping. Can we ever reach the store? If not, is there some small distance before the store that we can never cover? Can we ever go beyond the store? All these questions can be answered by showing our progress as an infinite series.

We write down how far we walk in each interval and then add all these distances together.

$$\textit{Distance We Walk} = \frac{1}{2} + \left(\frac{1}{2}\right)\cdot\left(\frac{1}{2}\right) + \left(\frac{1}{2}\right)\cdot\left(\frac{1}{4}\right) + \left(\frac{1}{2}\right)\cdot\left(\frac{1}{8}\right) + \ldots$$

With the above series we show that each term is generated by taking one half the remaining distance to the store. By multiplying the terms out and using the sigma form, we get the infinite series:

$$\textit{Distance We Walk} = \sum_{n=1}^{\infty}\left(\frac{1}{2^n}\right) = \frac{1}{2} + \frac{1}{4} + \frac{1}{8} + \frac{1}{16} + \ldots$$

Now, the question is: do we ever get to the store—do we ever reach the 1-mile-mark? Of course, we can't do this exercise in the physical world. If we tried we would become very frustrated when we drew exceedingly near to the store's front door. The little distances represented by each successive new term in the series would become so small that we could not shuffle our feet forward that tiny amount. We would be so close to the store that we would be tempted to declare that we had actually arrived.

What we are really doing in this exercise is performing the walk as a mental exercise. Can we mentally subdivide the distance from our home to the store into an infinite series of distances, such that each interval is exactly one-half of the previous one? This very idea confused the ancient Greeks and made them shy away from ideas involving infinity. But we know that an infinite set of mathematical objects can exist because we have already learned about the number sequence. We also know we can construct an infinite series of numbers (representing our distances). The question is: Does this infinite series add up to exactly one mile or does it add to something less, more, or maybe some indefinite amount we cannot assign a number to?

Can we get to the store in a finite number of steps? Suppose someone claimed that this were true. We would ask: how many walks are needed? Suppose they answered: ten walks. However, we can add the distance for ten walks and see how far that is: $1/2 + 1/4 + 1/8 + 1/16 + 1/32 + 1/64 + 1/128 + 1/256 + 1/512 + 1/1024 = 1023/1024$ of a mile. This is very close to one mile, but is not quite there, being just over 5 feet short. Our claimant may object, "I'm almost there, I just need to add a few more terms." He adds more terms, but alas, when we add these terms to what we already have we are still short of 1 mile. We can never arrive at the store in a finite number of walks.

If we can never reach the store, how close can we get? Now our antagonist changes his mind and claims that no matter how many walks we make, we will always stop some very small distance from the store—some ever so short distance. "Fine," we reply. "Just tell us what this small distance is." Suppose our antagonist says, "That small distance is 1/1,000,000 of a mile or 0.000001 mile." We can easily add together enough walks to get closer to the store than 0.000001 mile. If we make only twenty walks we get closer to the store than 1/1,000,000 of a mile, for

$$\sum_{n=1}^{20} \frac{1}{2^n} = 0.9999990463$$

Having demonstrated that our twenty walks brings us closer to one mile than 1/1,000,000 of a mile, our claimant now wishes to revise his number and make it much smaller. If he does, we can always add enough terms to get closer to the store than this new amount. It doesn't matter how small our claimant makes the distance, we can always get closer to the store than this small distance. However, in a finite number of walks we can never reach exactly one mile, while at the same time getting as close to 1 mile as we wish.

This is the very heart of the meaning of a limit. A limit is the number (or distance, in our case) that we can get as close to as we wish, but which we can never reach or exceed in a finite number of additions. Therefore, a limit is a number which acts as a very special boundary. It is a boundary because it represents a place we cannot reach, yet it is a special boundary since it is the one number that we can get as close to as we wish.

When we want to show the value of a limit we use the limit sign (lim). In our example we say that

$$\lim_{n\to\infty} \left(\sum_{j=1}^{n} \frac{1}{2^j} \right) = 1$$

You'll notice this is slightly different from our normal expression for an infinite series. Inside of the parentheses is the symbol for a finite series which begins when $j = 1$ and continues adding $1/2^j$ until j is equal to n. We have had to introduce a new indexing symbol, j, because we now need two such symbols, n and j. Both n and j are arbitrary choices, and any two distinct symbols will do.

Now the limit symbol (lim) says that if we let n increase, we will get successive finite series with increasing numbers of terms, and these series will get closer and closer to 1 without ever reaching it. Many students become confused with the limit sign, thinking it is a command to do something, such as adding an infinite number of terms. They object that one cannot carry out such an addition. This is not what the limit sign stands for. It only identifies a number which is the limit to all the possible finite series we could construct. Therefore, when we encounter the "lim" sign we should think of it as a number which limits some process, a process which we may never actually carry out to completion. The reason for defining a limit in such a fashion is to avoid relying on the notion of a mathematical idea depending upon an infinite number of operations. Our concept of limit avoids this since it is defined by two ideas:

1. For any *finite* number of terms the series fails to reach the limit.
2. Given any value, no matter how small, we can find a *finite* number of terms of the series that gets us closer to the limit value than that small value.

In neither step above did we require an infinite process. However, this does not mean that we have completely eliminated the idea of infinity from our definition of a limit. In step 1 we say "for any finite number." How many finite numbers are there? An infinite number, of course. In the second step we say "given any value." Again, we are talking about an infinite number of possible values. We see that while we can eliminate the idea of performing an infinite number of operations, we still are dealing with the idea of the infinite.

Because the above expression is so cumbersome, we generally just drop the limit sign and show the series as:

$$\sum_{j=1}^{\infty} \frac{1}{2^j} = 1$$

In this example we have not actually added up an infinity of anything, but have only discovered the boundary to all of the above series containing only a finite number of terms. In our loose way of speaking we sometimes say that the infinite series adds to 1, or that if we let the number of terms go to infinity, we get 1. But we should remember that a limit is just a number which has been defined as a boundary to a process.

Possibly the most important constant in mathematics, e (pronounced as the letter e), is defined as a limit value. While π was the first irrational number encountered by humans and is generally recognized by most people because of the basic geometry we all learned in elementary school, e turns out to be a number that is just as important as—or possibly more important than—π. It is a fabulous irrational number with many surprising properties. We cannot give the precise value of e in decimal or fraction form because it is an irrational number, but an approximate value of e is: $e = 2.718281828459045\ldots$

Pi has an approximate value of 3.14159 . . . but this is about as far as I ever remember its value because the digits have no pattern. While the decimal expression of e has no overall pattern (being irrational), the first digits do have a pattern that is easy to remember. We begin with 2.7, then we write the year 1828 twice. This is followed by 45, then 45 doubled (90) and 45 again. From these short memorized numbers (2.7, 1828, and 45) we can recreate the value of e up to fifteen places right of the decimal point, an accuracy for e far in excess of anything ever needed for applied problems.

The formal definition of e is very interesting and is intimately connected to compound interest. We use the following limit to define e.

$$\lim_{n\to\infty}\left(1+\frac{1}{n}\right)^n = e$$

If we compute $(1 + 1/n)^n$ using larger and larger integers for n, we get a value that is closer and closer to e. For example, trying the calculation for $n = 1, 2, 5$, and 100 yields approximate values for e of 2, 2.25, 2.48832, 2.7048138 . . . .

The exact value of e can be represented by an infinite series in addition to the above limit. For example, we have the following remarkable infinite series representation.

$$e = \sum_{n=0}^{\infty}\frac{1}{n!} = 1 + 1 + \frac{1}{2} + \frac{1}{6} + \frac{1}{24} + \ldots$$

The symbol "$n!$" stands for all the numbers from 1 through n multiplied together and is read as n factorial. Therefore, $2! = 1\cdot 2 = 2$, $3! = 1\cdot 2\cdot 3 = 6$, while $4! = 1\cdot 2\cdot 3\cdot 4 = 24$. For example, if we have a can of peaches, we have $1! = 1$. Now, if we put two cans of peaches in one box we have $2! = 2$ cans of peaches per box. If we combine three boxes of peaches into one case we get $3! = 6$ cans of peaches per case. Now we ship 4 cases of peaches and get $4! = 24$ cans of peaches per shipment. Factorial numbers increase rapidly as n increases, e.g. $10! = 3{,}628{,}800$. The above infinite series expression for e turns out to be one of the most important infinite series in mathematics, as we shall see later.

The limits to some expressions turn out to be utterly amazing. For example, we have already mentioned the Fibonacci sequence. Suppose we were to take successive pairs of terms from the sequence and build a fraction with the larger term on top. Since the sequence is 1, 1, 2, 3, 5, 8, 13, . . . we get the successive fractions: $1/1$, $2/1$, $3/2$, $5/3$, $8/5$, and $13/8$. If we compute the decimal representations to these fractions we get: 1.0, 2.0, 1.5, 1.66 . . .,

1.6, and 1.625. These values appear to be getting ever closer to some definite amount. Do these fractions have a limit, and what would it be? Now for the shocker. These fractions do have a limit and that limit is the Golden Mean! That's right—the Golden Mean discovered by the ancient Greeks, which has an exact value of $(1 + \sqrt{5})/2$, is the limit to fractions formed from successive Fibonacci terms.

$$\lim_{n=\infty} \frac{F_{n+1}}{F_n} = \phi = \frac{1 + \sqrt{5}}{2}$$

This is such an astounding and unexpected result. We ask ourselves in wonderment, how can the Fibonacci sequence be so intimately connected with something that appears to be entirely disconnected, the Golden Mean. This is the astonishing thing about mathematics. We are constantly finding connections between what appear to be completely separate entities. The hair on my head rises when I encounter such beautiful strangeness. It is as if there is some great magic going on just under the surface of reality which I cannot see, but see the results, instead. Remember when we were talking about the beauty of equations, we pointed out that the Golden Mean was exactly equal to the simplest infinite continuing fraction and also the simplest infinite nested radical. Now we suddenly realize that both these expressions must be exactly equal to the limit of the ratio of successive terms of the Fibonacci sequence!

However, we are not yet done with infinite series, for there are many to enjoy.

CONVERGING OR DIVERGING?

"The divergent series are the invention of the devil . . ."—Niels Henrik Abel (1802–1829)[4]

A major question regarding any infinite form, be it sequences or series, is whether the form converges to a finite limit (is convergent) or not (is divergent). Even today this can be a ticklish problem. We have already seen an example of a diverging series in Σn, and a converging series in $\Sigma(1/2^n)$. A famous series which was a puzzlement for many years is the harmonic series:

$$\textit{Harmonic Series} = \sum_{n=1}^{\infty} \frac{1}{n} = \frac{1}{1} + \frac{1}{2} + \frac{1}{3} + \ldots$$

Because the terms are always getting smaller, it is tempting to believe that the harmonic series converges. We can say that the limit of the individual terms is zero, as n goes to infinity. This means that for any small number we can find a term that is smaller than that number. The fact that the individual terms have zero as a limit does not ensure that the series (the addition of all the terms) is a fixed amount. It was Nicole Oresme (1323–1382), the Bishop of Lisieux, France, who gave a proof that the series diverges by rearranging the terms to demonstrate that the series contains an infinity of halves. This means that if we pick any number, however large, we can add enough terms in the harmonic series to exceed that large number. What is truly remarkable is that the harmonic series grows in size so slowly. In order for the harmonic series to reach just 10, we must add together its first 12,367 terms, and to reach 100 we must add together the first 1.5×10^{43} terms! We won't even consider how many must be added to reach 1000. But no matter how large of a number we choose, we can add enough terms of the series to exceed that number.

Are there any general principles we can use to determine if an infinite series is convergent? Many excellent books have been written on infinite series that address this very point. First, if we have a series whose terms are all identical, we can easily prove that such a series is divergent. Therefore, the terms must be decreasing, or alternate in sign, i.e., each successive term has a sign that is opposite the previous term, and the signs alternate between plus and minus.

Another historically interesting infinite series is the series of alternating ones:

$$\sum_{n=\infty}^{\infty} (-1)^n = 1 - 1 + 1 - 1 + 1 - 1 + 1 \ldots$$

For any finite number of terms, the series is either 0 or 1, depending on whether we consider an even number of terms (zero) or an odd number (1). If we consider only one term, the series is equal to 1. If we take the first two terms, $1 - 1$, then the series is equal to 0. Three terms again produce a 1, while four terms yield 0. Some mathematicians want to call it an indefinite series because it does not grow. Others want to call it divergent because it does not converge to a single value. Leibniz, the cofounder of calculus, thought that the limit of the series was ½. Most texts now call such a series divergent, because any series that is not convergent is designated as divergent.

A monotonic series is one that is either always increasing or always decreasing. For a monotonic decreasing series we have the condition that $a_{n+1} < a_n$ for all terms within the series. Such a decreasing series is also known as evanescent. It was once thought that if a series was evanescent, then it converged, but we have seen that the harmonic series is divergent, and it is definitely evanescent. How can we tell when an evanescent series is convergent? One test is called the comparison test. This is a difficult problem because for any evanescent series that is divergent, we can construct another that is also divergent, but diverges slower than the first. Hence, no comparison test will work for all series.

> *The Comparison Theorem*: If a series has all positive terms and converges, and if corresponding terms of this series are always larger than corresponding terms of a second series, then the second series also converges. If a series has all positive terms and diverges, and if the corresponding terms of this series are always smaller than the corresponding terms of a second series, then the second series also diverges.

Therefore, to demonstrate that a particular series converges all we have to do is find another series that converges and demonstrate that the corresponding terms of the first series are smaller than those of the converging series. When considering questions of convergence or divergence, we can actually disregard any finite number of terms and consider the series after these terms have been taken out. Thus, we can make the above theorem stronger by saying that the second series converges if its terms are smaller than those of the first series, beginning with some specific term of the series.

For example, the series consisting of the reciprocals of the primes diverges or:

$$\sum_{2}^{\infty} \frac{1}{p} = \frac{1}{2} + \frac{1}{3} + \frac{1}{5} + \frac{1}{7} + \frac{1}{11} + \ldots = \infty$$

What about the series consisting of the reciprocals of all odd numbers?

$$\sum_{n=1}^{\infty} \frac{1}{2n-1} = \frac{1}{1} + \frac{1}{3} + \frac{1}{5} + \frac{1}{7} + \frac{1}{9} + \ldots = ?$$

Comparing the first five terms of the two series we have:

term	$1/(2n-1)$		$1/p$
1	$1/1$	$>$	$1/2$
2	$1/3$	$=$	$1/3$
3	$1/5$	$=$	$1/5$
4	$1/7$	$=$	$1/7$
5	$1/9$	$>$	$1/11$

All of the terms of the odd reciprocal series are either greater than or equal to the corresponding terms of the reciprocal prime series. When we proceed beyond the fifth term, we realize that the denominators of the odd series will always be smaller than the corresponding prime series. Therefore, the reciprocals of the odd series will always be larger than those of the prime series. Since the prime series diverges, the odd series must also diverge.

We have just scratched the surface of this very fascinating subject, and other kinds of tests are available for infinite series.[5] But we must go on.

MOST WONDROUS SERIES

We are now prepared to look at some fascinating infinite series. We have already mentioned the harmonic series, which is just the sum of the reciprocals of the number sequence. And we know that both the series of the reciprocals of odd numbers and the reciprocals of primes also diverge. We can see that both of these series must diverge slower than the harmonic series. Yet, given any diverging series, we can find one that diverges slower—just subtract the first n terms.

But what about series that converge? One of the most famous is the sum of the reciprocals of square numbers:

$$\sum_{n=1}^{\infty} \frac{1}{n^2} = \frac{1}{1^2} + \frac{1}{2^2} + \frac{1}{3^2} + \ldots = \frac{\pi^2}{6}$$

But wait! The sum of this series involves π. How did π get into an infinite series? π is the ratio of a circle's circumference to its diameter; why is it now appearing in an apparently noncircle way? This, again, is the amazing thing about mathematics. Just when we think we're getting it all figured out, something strange and unexpected jumps up before us.

But the above is not the only series where π shows up, for we have:

$$\sum_{n=1}^{\infty} \frac{1}{n^4} = \frac{1}{1^4} + \frac{1}{2^4} + \frac{1}{3^4} + \frac{1}{4^4} + \ldots = \frac{\pi^4}{90}$$

Here again we encounter π. Since both series involve π raised to the same power as the exponent in the denominator of each term, is there a connection? The value of the series $\Sigma\, 1/n^2$ was discovered by Leonhard Euler (1707–1783), one of the most prolific mathematicians to have ever lived (Figure 58). His works if published in one set of volumes would fill over ninety books. Euler loved infinite series and discovered many wonderful things about them. Beside determining the value of $\Sigma\, 1/n^2$ he also evaluated the infinite sums where the exponents are even numbers or $\Sigma\, 1/n^{2k}$ for

FIGURE 58. Leonhard Euler (1707–1783). Photograph from Brown Brothers, Sterling, PA.

every $k \geqslant 1$. He accomplished this by making use of the strange and wonderful Bernoulli numbers, which themselves are the coefficients to another infinite series. Using Bernoulli numbers, Euler showed that the series with even exponents could be written as the following:

$$\sum_{n=1}^{\infty} \frac{1}{n^{2k}} = (-1)^{k+1} \frac{(2\pi)^{2k} \cdot B_{2k}}{2(2k)!}$$

where the B_{2k} is the $2k$ Bernoulli number. The first few Bernoulli numbers are $B_0 = 1, B_1 = 1/2, B_2 = 1/6, B_3 = 0$, and $B_4 = -1/30, B_5 = 0$, and $B_6 = 1/42$. If we let $2k = 4$, then $k = 2$ and $B_{2k} = -1/30$. Substituting this into the above equation yields $\Sigma 1/n^4 = \pi^4/90$. When we let $2k = 6$, then $k = 3$ and $B_{2k} = 1/42$. Substituting this value into the above equation we get $\Sigma 1/n^6 = \pi^6/945$. Euler applied the same kind of analysis to the sum of reciprocals of odd numbers raised to even powers. For example:

$$\sum_{n=1}^{\infty} \frac{1}{(2n-1)^2} = \frac{1}{1^2} + \frac{1}{3^2} + \frac{1}{5^2} + \ldots = \frac{\pi^2}{8}$$

and,

$$\sum_{n=1}^{\infty} \frac{1}{(2n-1)^4} = \frac{1}{1^4} + \frac{1}{3^4} + \frac{1}{5^4} + \ldots = \frac{\pi^4}{96}$$

There are many more infinite series involving the constant π. For example we have the very nice series discovered by James Gregory (1638–1675), a brilliant Scottish mathematician who died at the early age of thirty-six.

$$\sum_{n=1}^{\infty} \frac{(-1)^{n+1}}{2n-1} = 1 - \frac{1}{3} + \frac{1}{5} - \frac{1}{7} + \frac{1}{9} - \frac{1}{11} + \ldots = \frac{\pi}{4}$$

What a beautiful infinite series!

We call a converging infinite series that alternates in sign between plus and minus either an absolutely convergent series or a conditionally convergent series. A series is absolutely convergent if the same series with all positive signs is also convergent. For example, the series

$$\sum_{n=1}^{\infty} \frac{(-1)^{n-1}}{n^2} = 1 - \frac{1}{4} + \frac{1}{9} - \frac{1}{16} + \ldots$$

is absolutely convergent since the associated series,

$$\sum_{n=1}^{\infty} \frac{1}{n^2} = 1 + \frac{1}{4} + \frac{1}{9} + \frac{1}{16} + \ldots = \frac{\pi}{6}$$

where the negative signs are replaced with positive signs, is convergent. However, the series

$$\sum_{n=1}^{\infty} \frac{(-1)^{n-1}}{n} = 1 - \frac{1}{2} + \frac{1}{3} - \frac{1}{4} + \ldots$$

is only conditionally convergent since the associated series

$$\sum_{n=1}^{\infty} \frac{1}{n} = 1 + \frac{1}{2} + \frac{1}{3} + \frac{1}{4} + \ldots$$

is the harmonic series and is not convergent.

We must use caution when working with infinite series with alternating terms because we can rearrange the terms in different ways so that the series converges to different values, or even diverges. Therefore, when working with infinite series with alternating signs we must be careful and remember that the order of the terms is most important.

We can change the pattern of the signs in Gregory's series to terms alternating in pairs of plus signs and minus signs to get:

$$\sum_{n=1}^{\infty} (-1)^{n+1}\left(\frac{1}{4n-3} + \frac{1}{4n-1}\right) = 1 + \frac{1}{3} - \frac{1}{5} + \frac{1}{7} - \frac{1}{9} + \frac{1}{11} - \ldots \frac{\pi}{4}\sqrt{2}$$

Another nice alternating series is due to Euler again.

$$\sum_{n=1}^{\infty} \frac{(-1)^{n+1}}{(2n-1)^3} = \frac{1}{1^3} - \frac{1}{3^3} + \frac{1}{5^3} - \frac{1}{7^3} + \ldots \frac{\pi^3}{32}$$

Srinivasa Ramanujan Iyangar (1887–1920), known simply as Ramanujan, was a poor young Indian mathematician, who had virtually taught himself. The famous British mathematician G. H. Hardy recognized Ramanujan's genius and enticed him to Trinity College, Oxford, England. Unfortunately Ramanujan contracted a disease, probably tuberculosis, and died after returning to India. The young genius had the ability to peer deeply into the wellspring of mathematics and snatch out beautiful expressions. During his short life he discovered the sums of many infinite series. The following is one of his rather complex infinite series.

$$\frac{\sqrt{8}}{9801} \sum_{n=0}^{\infty} \frac{(4n)!}{(n!)^4} \cdot \frac{1103 + 26390n}{396^{4n}} = \frac{1}{\pi}$$

Does every converging infinite series converge to a value involving π? Of course not. Consider the following, also the work of Ramanujan.

$$\sum_{n=0}^{\infty} (-1)^n \frac{(2n+1)^3 + (2n+1)^2}{n!} = \frac{1^3 + 1^2}{0!} - \frac{3^3 + 3^2}{1!} + \frac{5^3 + 5^2}{2!} - \ldots = 0$$

We could go on indefinitely looking at interesting infinite series, but we must stop somewhere. For now let's consider the sum of the reciprocals of the triangular numbers, which were first discovered by the Greeks as the sum of n terms of the counting sequence, e.g., $1 = 1$, $3 = 1 + 2$, $6 = 1 + 2 + 3$, and $10 = 1 + 2 + 3 + 4$, etc. The reciprocals of the triangular numbers form the sequence, $\frac{1}{1}, \frac{1}{3}, \frac{1}{6}, \frac{1}{10}, \ldots$. Gottfied Leibniz (1646–1716) found the sum for this series:[6]

$$\sum_{n=1}^{\infty} \frac{2}{n(n+1)} = \frac{1}{1} + \frac{1}{3} + \frac{1}{6} + \frac{1}{10} + \frac{1}{15} + \ldots = 2$$

In the following chapters we will introduce more infinite series connected with specific functions, and we will see that such series are an integral part of all mathematics—especially calculus. Yet, we should pause for a moment to consider an astonishing discovery we have made. Certain numbers exist, such as π and e, which seem to appear much more frequently than we should expect. Why? Suppose we went to the Metropolitan Opera one evening and sat beside a lovely woman wearing a red dress and a ruby necklace. The next day we go to the New York Stock Exchange, and there on the floor of the Exchange is the same woman, now wearing a black dress and diamonds. Next we have lunch at the Ritz, and behold, there she is again, this time in a green dress and emeralds. Would this make us wonder what was going on, as if that woman played some very important role either in our own lives or in the life of New York City?

It is this way with the constants π and e, for they keep appearing where we don't expect them. Is there some enigmatic connection underlying mathematics, in a kind of metamathematics which we don't yet understand? There are so many mysteries yet to solve!

6

Functions

So far we have studied equations that involve two unknowns, x and y, which have been either linear or raised to the second power (quadratic). In mathematics there are, of course, many other equations graced by two unknowns, x and y. Here, we will explore some of these equations known as elementary functions.

Let us begin by defining what we mean by a function. Mathematics is fundamentally about relationships between different sets of objects, and a function is a special kind of relationship between such sets. In most cases the objects within the sets are real numbers, which fill up the real number line, or complex numbers (numbers we will introduce shortly). Let's see an example of a very simple relationship between two sets of numbers where each set has just three members (elements).

$$\begin{array}{c} \{\,1,\, 2,\, 3\,\} \\ \downarrow\ \downarrow\ \downarrow \\ \{\,1,\, 4,\, 9\,\} \end{array}$$

As is the custom, we have enclosed each set with special brackets to identify it as a set. The first set consists of the numbers 1, 2, and 3, while the second set consists of the numbers 1, 4, and 9. We have assigned each number in the first set to a specific number in the second set. You can see at once how this assignment was carried out. The numbers in the second set are just the squares of those in the first. This particular assignment was done just for convenience. In reality, we can assign any number from the first set to any number in the second. Therefore, another relationship between these same sets could be defined as:

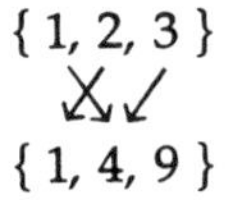

$$\begin{array}{c} \{\,1,\, 2,\, 3\,\} \\ \searrow\!\!\!\!\swarrow\ \swarrow \\ \{\,1,\, 4,\, 9\,\} \end{array}$$

Here, we have assigned the number 1 in the first set to 4 in the second, 2 is assigned to 1, while 3 is also assigned to 4. Nothing is assigned to 9. This demonstrates that two objects from the first set can be assigned to one object in the second, and that objects can be in either set that are not assigned to anything. Connecting members of one set to those of another set is called mapping. When we assign one and only one element from each set to the other, as we did in the first example, then we have a one-to-one mapping. When we count a set of objects, we use one-to-one mapping to assign each successive number word to one object within the set we wish to count. In the second example above we assigned two different elements of the first set to one in the second set. This is called a many-to-one mapping. There can also be a one-to-many mapping.

Now we are going to jump up a level of complexity and consider the simple equation $y = x$. This is the equation of a straight line that passes through the origin of a Cartesian coordinate system, as we noted earlier, and is called the identity line (Figure 59). If we let the elements of our first set be all real numbers on the x-axis and the elements of the second set be all the real numbers on the y-axis, then we see that this equation, $y = x$, defines a mapping between two infinite sets of real numbers, namely all the real numbers on the x-axis mapped onto all the real numbers on the y-axis. In addition, the mapping is one-to-one, for every x has a unique y,

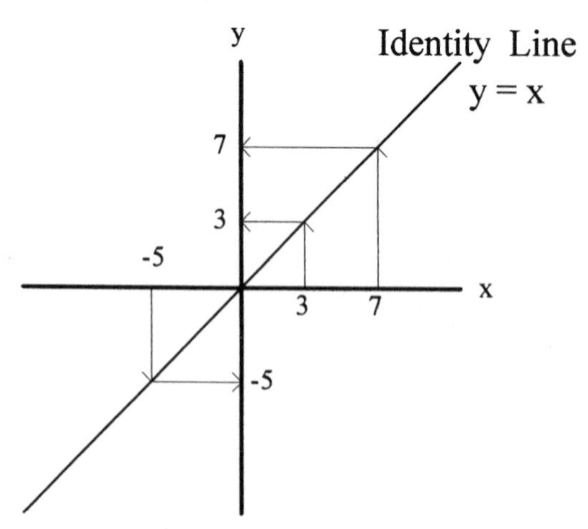

FIGURE 59. The identity line, $y = x$, defines a mapping of the real numbers of the x-axis onto the real numbers of the y-axis.

and every y a unique x. In other words, each and every real number along the x-axis has been mapped to a unique real number on the y-axis; no y-axis number is assigned to two x-axis numbers, and no x-axis number is assigned to two y-axis numbers. This example demonstrates that sets involved in mapping can be both one-to-one and infinite. Now we are ready to define a function.

FUNCTION: a mapping between two sets that is not one-to-many.

The only restriction we place on a mapping to be a function is that we can't have one element of the first set mapped to more than one element in the second set. Our first set is called the *domain* of the function, while the second set is called the function's *range*. For a relationship to be a function, we can define any kind of mapping that we want as long as no element of the domain maps to two or more elements in the range.

Sometimes it is instructive to look at sets that contain elements other than numbers. Let's take the set of all biological mothers as our domain and the set of all biological daughters as the range. Do we have a function in the mother–daughter relationship? Some mothers in the domain will have more than one daughter in the range. This violates the definition of a function, so this relationship is not a function. However, if we let the daughters be the domain and map them onto their mothers in the range then we do get a function, because no daughter will be mapped to two different biological mothers. However, more than one daughter may map to one mother (many-to-one mapping). In mathematics we often define mappings between sets whose elements are not numbers. For example, we might map geometric spaces or logical propositions. We can even have a domain that consists of functions mapped to a range made up of other functions. Let's recap what we have in terms of mappings and functions.

A one-to-one mapping = a function
A many-to-one mapping = a function
A one-to-many mapping = not a function

To demonstrate the concept of a function, let's look at the conic sections and decide if they are functions or not. When we graph our relationships, we will graph the domain from the x-axis and the range from the y-axis. First, look at the ellipse in Figure 60, which has the algebraic equation, $x^2/9 + y^2/4 = 1$. If we travel along the x-axis to the point x_1 we

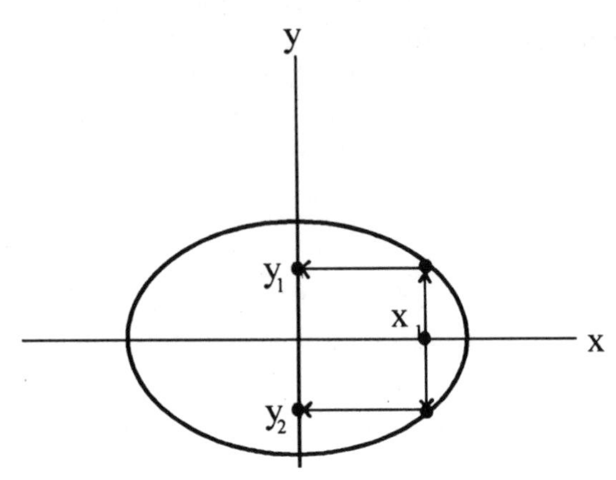

FIGURE 60. The ellipse is not a true function because the point x_1 maps onto both y_1 and y_2.

can see that this x value is associated with the two y values of y_1 and y_2. This means that an ellipse is not a function because every x between the ellipse's vertices maps to two different y values, violating the definition of a function.

Now look at the parabola in Figure 61a which has the algebraic equation of $y = x^2$. Every x along the x-axis is mapped to only one value of y, no matter how far from the origin we move. Hence, $y = x^2$ is a function. We can understand this in the algebraic sense: if we substitute a value for x in the parabola's equation, we get only one value for y. If we rotate this parabola 90 degrees we get the parabola $x = y^2$ shown in Figure 61b. We see at once that it is not a function. If we solve this equation for y we get $y = \pm\sqrt{x}$. Now, if we substitute a value in for x, we generate two different y values. Hence, just being a parabola does not guarantee we have a function.

How do we tell if a relationship is a function? This is not always easy to do. One test is to look at the graph of the relationship and see if there exists a vertical line that hits the graph in two or more places. If so, then the relationship cannot be a function. This is called the vertical line test. Of course, just by looking at a graph we can't always tell if a vertical line will touch the graph twice. What we really need is a method that determines whether any particular value of x generates more than one value for y. In many cases when we introduce an algebraic relationship we will be able to determine at once if it is a function or not.

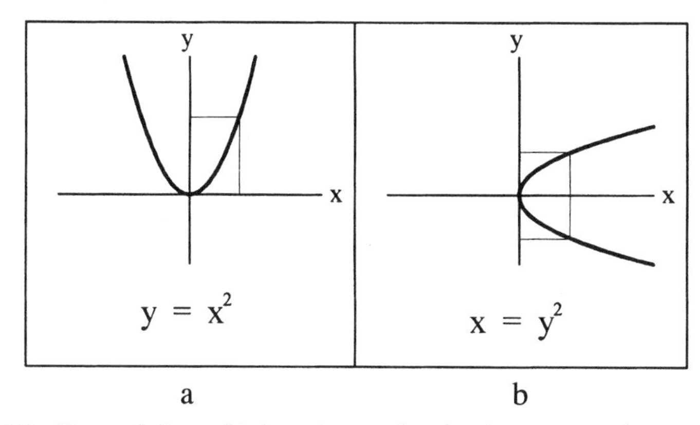

FIGURE 61. The parabola $y = x^2$ in box a is a true function since every x value maps to only one y value. However, the parabola $x = y^2$ in box b is not a function because all positive x values map to two y values.

A notation has evolved which we use to indicate when we are dealing with a function. If we wish to indicate that we have a functional relationship between y and x we write it as $y = f(x)$. This notation takes a little getting used to. The $f(x)$ does not mean that a number f is multiplied by another number x. What it does mean is that we are choosing to call our relationship f and that the unknown associated with y is x. Therefore, instead of saying that y is related to x in a functional relationship (which uses too many words) we simply write $y = f(x)$, and this says the same thing. If we wish, we can define a second functional relationship between x and y. However, we can't use f again, so we choose another letter from the alphabet, say g, and write $y = g(x)$. Now we have two functions, each defined differently in terms of x, and we can discuss them without getting confused. This type of notation for functions was first used in 1735 by Leonhard Euler in his *Commentaries*.[1]

Let's take an example. Suppose we have the function $y = 2x - 1$. Because $y = f(x)$, we can also write it as $f(x) = 2x - 1$. The f stands for the rule that associates x with y. In this case the rule is: multiply x by 2 and subtract 1. We can define a second function for y as: $y = 3x^2 + 9$. We don't want to call this new rule f because, in this example, we already have a rule defined for f, i.e., $2x - 1$. We need a new letter, and we'll choose g this time: $y = g(x)$ where g is the rule that says to get y, square x, multiply the result by 3, and add 9.

POLYNOMIAL FUNCTIONS

The very simplest polynomial functions are just constant functions which we can represent as $f(x) = C$, where C stands for some real number. Using the y-axis in a Cartesian coordinate system to represent the values of $f(x)$ and the x-axis to represent the x values, our functions graph nicely. The graph of function $f(x) = C$ is a horizontal line that passes through the y-axis at point C. Using the definition of function, we can categorize functions into different kinds. The next higher level functions are the polynomial functions that involve some unknown, x. Every such polynomial is of the form:

$$f(x) = a_n x^n + a_{n-1} x^{n-1} + \ldots + a_1 x + a_0$$

and meets the definition of a function. We can see that such equations are functions because no matter what value of x we substitute into the equation, we generate only one value for $f(x)$. This means that the polynomial equations

$$y = 3x + 4$$
$$y = 2x^2 - 9x + 14$$
$$y = -x^3 + 12x^2 - 4x + 1$$

are all functions. In addition, polynomial equations are very well behaved functions. Not only are they all functions, but they are continuous everywhere. This means that if a polynomial graph were a road, we could drive on that road from any point on the graph to any other point on the graph. There simply are no gaps in the graph. Plus, the road never has sharp turns (cusps) but is always smooth (Figure 62). Just the kind of road on which to go motorcycle riding on a sunny afternoon. The feature of being smooth everywhere is very important to calculus.

We can use polynomial functions to generate the next level of elementary functions, the rational expressions. A rational expression is simply a fraction where both the numerator and denominator are polynomials. For example, a rational expression is:

$$\frac{x^2 - 2x - 15}{x^3 - 5x^2 + 2x + 8}$$

Figure 63 shows the graph of this rational expression. The dotted vertical lines in Figure 63 are not part of the curve, but are vertical asymp-

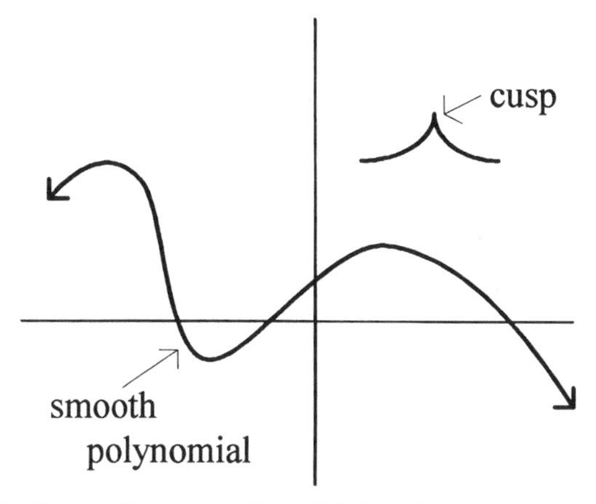

FIGURE 62. All polynomials are smooth, well-behaved functions without any breaks or cusps.

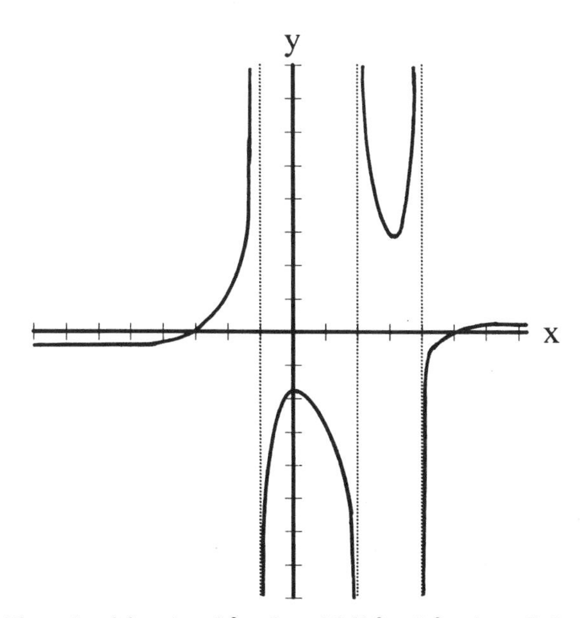

FIGURE 63. The rational function, $(x^2 - 2x - 15)/(x^3 - 5x^2 + 2x + 8)$, has four separate branches and four asymptotic lines, including the x-axis.

totes, or lines that our curves approach but never touch. In arithmetic, we are never allowed to divide a number by zero, for the result is undefined. We treat fractions made up of polynomials in the same fashion and never divide by zero. Therefore, we must exclude those values from the domain of our function (i.e., possible values of x) that make the denominator of our rational expression equal to zero. In the above case, the values of x which make the denominator equal to zero are $x = -1$, 2, and 4, for these three values are the three solutions to the polynomial $x^3 - 5x^2 + 2x + 8 = 0$.

You can see at once that rational expressions are not well behaved like their parents, the polynomials. Graphs of polynomial functions are connected, but the graph in Figure 63 is broken into four different branches. If you were riding a motorcycle on one branch, you could never hope to reach one of the other branches. Yet, the graph is both beautiful and interesting precisely because it is unconnected and because the branches race outward, dashing ever closer to their asymptotic lines. Rational expressions can have three kinds of asymptotic lines: vertical, horizontal, and oblique. An oblique asymptote is a line that is neither vertical nor horizontal. Figure 63 has three vertical asymptotes but it also has the x-axis as a horizontal asymptote.

We can get a rational expression with an oblique asymptote just by considering the reciprocal of our function in Figure 63.

$$\frac{x^3 - 5x^2 + 2x + 8}{x^2 - 2x - 15}$$

This rational expression is graphed in Figure 64. We can see that for extreme negative and positive values of x, our curve is approaching the line $y = x - 3$, which is our oblique asymptotic line.

We can do more with polynomials than simply divide them to generate rational functions. We can add and subtract them, multiply and divide them, and we can also raise them to powers and extract roots. For example, we can take the square root of the simple polynomial $x + 1$ to get the function $y = \sqrt{(x + 1)}$. We can employ the concepts of powers and roots to get more complex functions such as $y = (x^3 - 2x + 4)^{2/3}$. All such functions made from polynomials using ordinary algebraic operations are called algebraic functions, as opposed to the transcendental functions which we are about to discuss.

Functional notation is very handy when we work on applied problems. We can write that the volume of a right cylinder (the shape of an

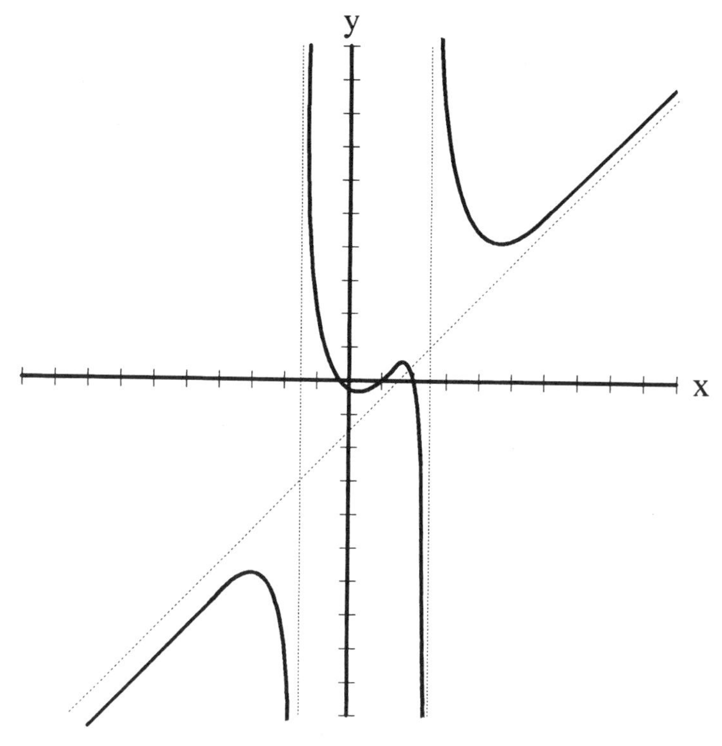

FIGURE 64. The rational function, $(x^3 - 5x^2 + 2x + 8)/(x^2 - 2x - 15)$ has two vertical asymptotic lines and one oblique asymptotic line, $y = x - 3$.

ordinary tin can) is a function of the cylinder's height and radius, or Volume = $V(h,r)$. Here, the rule is identified as V, and instead of x inside the parentheses, we have h for height and r for radius. We can show this equation as $V(h,r) = \pi r^2 h$. This says that the volume of a cylinder, V, is π times the square of the radius, multiplied by the cylinder's height.

Why all this bother about a seemingly obscure definition about mappings and functions? The study of continuity is called analysis, and is the door leading to almost all higher mathematics. Functions have very special features that allow us to do creative things in analysis. Knowing that a relationship is a function allows us to move into the field of calculus and beyond. For now, the reason to learn about functions is that we are going to have some fun with them.

PLAYING WITH FUNCTIONS

When we were dealing with the conic sections, we graphed various curves and then, by changing the equations, moved them around and changed their shapes. We can do the same with functions in general. Suppose we have graphed a function $y = f(x)$. Now I'm not telling you just what this function is or what its graph looks like, but we can easily move the entire graph in an upward direction by simply adding a constant onto the function or $y = f(x) + c$. In the same way we can move the entire graph of the function down by subtracting a constant. In fact, we can move functions about on the graph in many different ways. To illustrate these alterations, we will use the function $y = \sqrt{x}$ where x can have only positive values or zero. We have restricted the domain (possible values of x) since at this time we do not have any meaning when the number under the radical sign is negative. Figure 65a shows our graph of $y = \sqrt{x}$.

1. Shape: We can change the shape of a function by multiplying by a constant or $y = c \cdot f(x)$. Figure 65b shows the graph of $5 \cdot \sqrt{x}$, which is the basic graph of $y = \sqrt{x}$ but stretching up and to the right at a much faster rate.

2. Shift up and down: Here we simply add or subtract a constant to the function: $y = f(x) \pm c$. Figure 65c is the graph of $y = \sqrt{x} + 3$, which is the original graph of $y = \sqrt{x}$ but every point has been moved up three units.

3. Shift right or left: Substitute $x \pm$ c for x, remembering a negative c moves us toward the right, while a positive c moves toward the left. Figure 65d is the graph of $y = \sqrt{(x - 2)}$, which is the original graph moved two units to the right.

4. Reflect the graph around the x-axis: In this case we simply take the negative of the function or $y = -f(x)$. Figure 65e is the graph of $y = -\sqrt{x}$, which is the reflection of $y = \sqrt{x}$ around the x-axis.

5. Reflect the graph around the y-axis: Substitute $-x$ for x or $y = f(-x)$. Figure 65f is $y = \sqrt{(-x)}$, which reflects $y = \sqrt{x}$ around the y-axis. Notice that the new function, $y = \sqrt{(-x)}$, is not defined when x is positive, and that no part of the graph appears on the right side of the origin. The new function is defined only when x is negative, corresponding to a graph appearing only on the negative side of the x-axis.

Knowing how to alter basic functions allows us to visualize new algebraic expressions, because we can recognize how the simpler function has been changed by one of the five methods above. This means we can

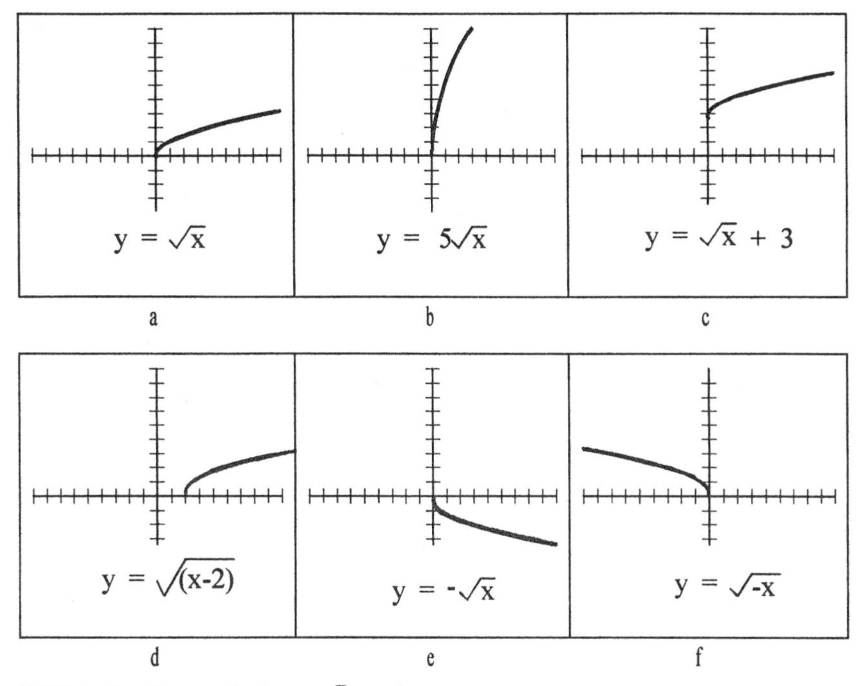

FIGURE 65. The graph of $y = \sqrt{x}$ can be moved about the Cartesian plane by making specific adjustments.

learn the basic forms of functions, and then use this information to visualize more complex expressions. For example, if we run into the function $y = -\sqrt{(x + 2)} - 3$ we need not be confused as to what the graph looks like. We start with the basic function $y = \sqrt{x}$, which we are now familiar with, and notice that the new function is reflected around the x-axis, moved two units left and three units down or:

$$\begin{array}{cc} \text{negative} & -3 \text{ moves} \\ \text{reflects} & \text{three down} \\ \uparrow & \uparrow \end{array}$$

$$-\sqrt{(x + 2)} - 3$$

$$\begin{array}{c} \downarrow \\ 2 \text{ moves} \\ \text{two to the left} \end{array}$$

What we are essentially doing is learning how to interpret a rather complex looking function by considering the simple function it is based upon, and actually visualizing how this function must look when it is graphed. With practice, this approach soon gives us the "taste" or "feel" for graphs of mathematical functions.

EXPOUNDING THE EXPONENTIAL

We are now ready for our next class of elementary functions. Most of us learned about exponents in high school algebra. Exponents are just those little superscripts written on numbers that instruct us to multiply a number by itself that many times. Hence, a^3 stands for $a \cdot a \cdot a$, and 3^5 stands for $3 \cdot 3 \cdot 3 \cdot 3 \cdot 3$. Gottfried Wilhelm Leibniz (1646–1716) codeveloped calculus along with Isaac Newton. In addition to his numerous achievements in both mathematics and philosophy, Leibniz made remarkable advances in notation, spending years in research while also conferring with numerous other mathematicians. In 1675, in his *Entdeckung der höheren Analysis*, he was the first to use the superscript number to indicate an exponent.[2] Before his time, it was the general rule to simply list the number of times a term was to be multiplied—a very cumbersome method.

When the exponent to a number, x, is 1 we have x^1, which we define as just x by itself. When the exponent is 0 we define x^0 as equal to 1. This defines exponents for all positive whole numbers and zero. How about negative whole numbers? For that we have the simple rule that $x^{-n} = 1/x^n$, also a notation introduced by Leibniz in 1684. Hence, we define negative whole number exponents as the reciprocals of the positive ones. For example, $3^{-2} = 1/3^2 = 1/9$, and $5^{-3} = 1/5^3 = 1/125$.

What about fractions such as ½? Can they be exponents? Yes. We achieve this by defining the roots of numbers. Most of us know that the square root of a number is another number which when multiplied by itself gives us the original number back. For example, 2 is the square root of 4 because $2 \cdot 2 = 4$. Therefore, if x is our number then $\sqrt{x}$ is the square root of x. This means that $\sqrt{x} \cdot \sqrt{x} = x$. We can also define the cube root of a number as another number that when multiplied by itself three times yields the original number: $\sqrt[3]{x}$ is the cube root of x and $\sqrt[3]{x} \cdot \sqrt[3]{x} \cdot \sqrt[3]{x} = x$. In such a manner we can define the nth root of a number where n is any positive whole number. Now we do a nice trick; we define the denominator of a fractional exponent to be the root of the base number. This means that $\sqrt{x} = x^{1/2}$ and $\sqrt[3]{x} = x^{1/3}$, and so forth. Now we can combine the powers and the roots. For example $x^{3/5}$ stands for the number x cubed and then we find the fifth root of the result. Or we can achieve the same end by finding the fifth root of x and then multiplying that number by itself three times. Combining powers and roots allows us to define exponents for any fraction p/q, where p and q are whole numbers (excluding $q = 0$).

Through our notation and definitions we have defined exponents for all rational numbers (fractions), both positive and negative. What about an exponent that is not rational? How about 2^{π}? Does it make any sense to raise 2 to the π power? Using the concept of limits we can actually define the meaning of 2^{π} and other irrational numbers used in exponents. Therefore, we can now talk about exponents taking the value of any real number, either rational or irrational. This is great, because it allows us to define a new function, the exponential function. An example of a simple exponential function is 3^x. What this means is that we are going to take the number 3 as a base and raise it to the appropriate power for every real number x. In functional notation we have $y = f(x) = 3^x$. Hence, for every value of x we get a corresponding value for $f(x)$. When $x = 1$ we get $f(x) = 3^1 = 3$, and when $x = 2$, $f(x) = 3^2 = 9$. To account for all real numbers, including zero, we define any base raised to the zero power as equal to 1. This means that $3^0 = 1$.

The graph of $y = 3^x$ is presented in Figure 66. Notice that the graph begins on the left (for negative x) very close to the x-axis and then increases until it crosses the y-axis at $y = 1$. This makes sense since we've already defined any base raised to the exponent of zero to be just the number 1. Moving toward the right, the graph of $y = 3^x$ continues up in an ever steeper curve. The curve we have for $y = 3^x$ is different from a circle, or any of the conic curves—the ellipse, parabola, or hyperbola. Hence, the exponential curve is an entirely new curve. We can create a different exponential curve by using different bases. Figure 67 shows the graphs of 2^x, π^x, 10^x. Notice how much steeper the 10^x curve is than the 2^x curve. This demonstrates that the larger the base number, the steeper the resulting graph. We have also included the graph of π^x to show that we can also use bases that are not rational numbers, but irrational, such as π and e.

We must pause for a moment to mention a special base. We have already introduced a fabulous constant, the number e. We can use e as the base to an exponential function or $y = e^x$, which has been graphed in Figure 68. This particular exponential function turns out to be so important that it

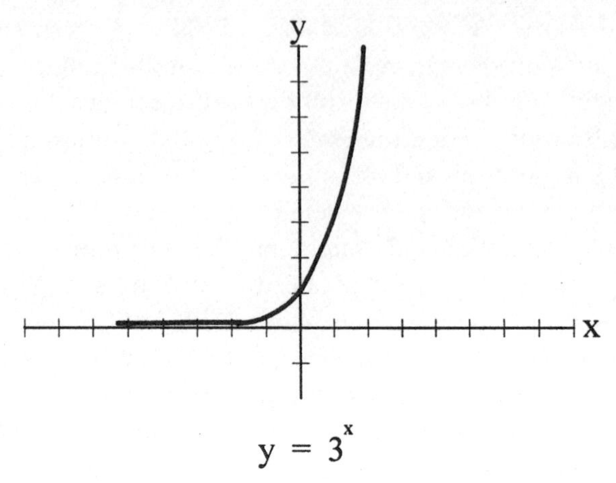

FIGURE 66. The graph of the exponential function $y = 3^x$. The graph approaches the x-axis as an asymptotic line on the left, crosses the y-axis at $y = 1$, and quickly grows in the first quadrant.

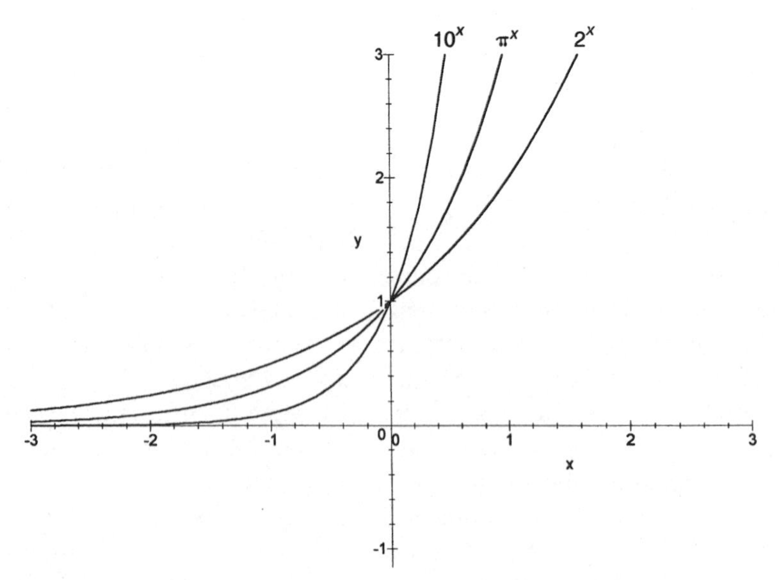

FIGURE 67. The graphs of the exponential functions 10^x, π^x, and 2^x. The larger the base, the quicker the graph climbs.

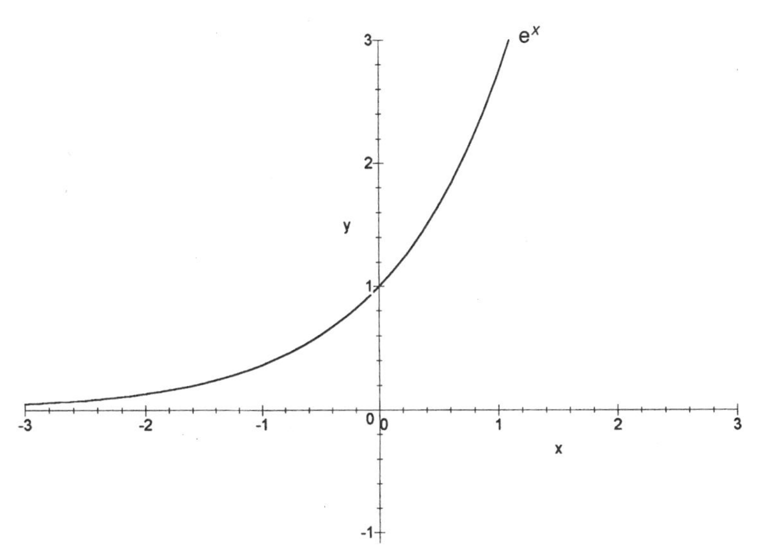

FIGURE 68. The graph of The Exponential Function $y = e^x$, one of the most important functions in mathematics.

is sometimes referred to as "The Exponential Function" instead of just "an exponential function." From now on, whenever I speak of the exponential function with the base of e, I'll use capital letters as in The Exponential Function, but when I am referring to an exponential function with another base number, I'll use lowercase letters. We have an excellent infinite series expression for The Exponential Function, which we will see is very useful:

$$e^x = \sum_{n=0}^{\infty} \frac{x^2}{n!} = 1 + x + \frac{x^2}{2} + \frac{x^3}{6} + \frac{x^4}{24} + \dots$$

This is precisely the infinite series we used to represent e, except we have added the x^n term in each numerator to account for the x term as an exponent to e.

The graph of The Exponential Function gives an excellent description of the natural growth of many processes found in the physical world. Populations of living organisms, from cells to human populations, tend to follow the Exponential curve when the x-axis is plotted as time and the y-axis is plotted as the population's size. This is especially true when populations have unlimited space and food, such as the early growth of cells in petri dishes, or when humans have historically moved into previously unoccupied lands.

Using our ideas of moving functions about, we can reflect The Exponential Function around the y-axis by simply putting a negative sign before the x so that we get $y = e^{-x}$, which is plotted in Figure 69. This curve also describes many natural processes including the loss of heat from a body when placed in cold surroundings, and the decay of radioactivity in radioactive substances. Therefore, The Exponential Function is used extensively in various fields of science to model many natural processes. It is no accident that The Exponential Function gives an accurate description of natural phenomena. When we take a look at calculus, we will come to understand why this is so.

We have already mentioned that some functions are one-to-one mappings, while others are many-to-one mappings. The parabola, $y = x^2$ is a many-to-one function because two different xs are mapped onto the same y value. For example, both $x = 2$ and $x = -2$ are mapped onto a y value of 4 since $4 = (2)^2$ and $4 = (-2)^2$. However, an exponential function ($y = a^x$) is a one-to-one mapping. This means that each and every x value is mapped to only one y value, and conversely each and every y value is mapped to only

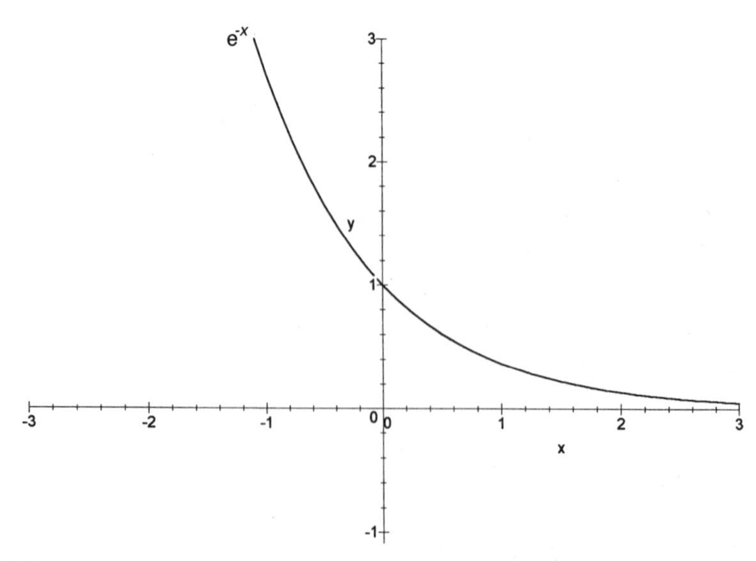

FIGURE 69. The graph of $y = e^{-x}$, The Exponential Function reflected around the y-axis.

one x value. Remember when we were testing a graph to see if it was a function? We introduced the vertical line test, which says that if a vertical line intersects with a graph in more than one point, then the relationship depicted by the graph is not a function. We can define a horizontal line test to test a graph for being a one-to-one function. In other words, when a horizontal line hits a graph in two places, then we know that a particular y is mapped to those two corresponding places on the x-axis, and the function cannot be one-to-one.

Now that we have done all the hard work to define both functions and one-to-one functions, we can take our reward. What we do now with one-to-one functions is great fun. We are going to interchange the x and y terms and find out what happens to the graph of the original function. Let's begin with the graph of the parabola $y = x^2$ where we are only going to allow the values of x to be positive. Figure 70 shows the graph of this

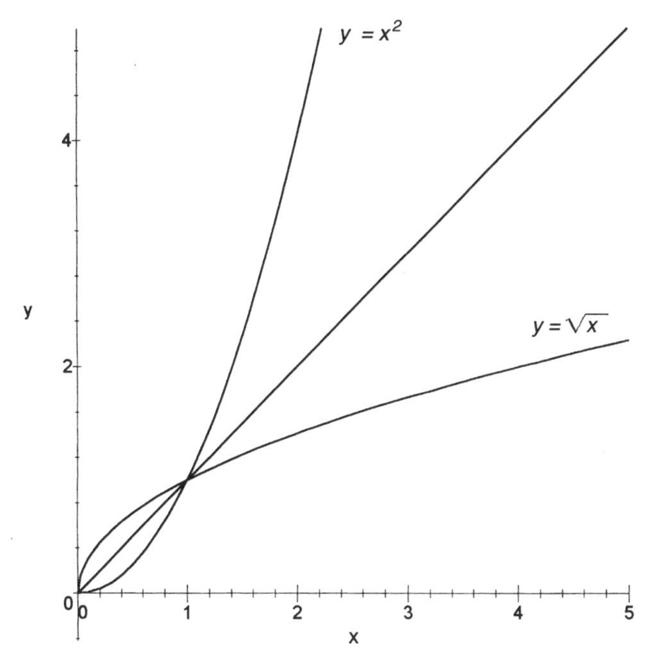

FIGURE 70. The reflection of the graph $y = x^2$ (positive values of x only) around the identity line to become the new function $y = \sqrt{x}$. Each of these two functions is the inverse function of the other. Every one-to-one function has its inverse function.

curve, which has no values to the left of the y-axis. We restrict the values of x because we only want to consider one-to-one functions, and if we allowed for x to be negative, then we would have an ordinary parabola, which is not one-to-one.

We take our restricted function, $y = x^2$, and swap the x and y terms, getting $x = y^2$. Now we solve for y, or $y = \sqrt{x}$. We have a new function which has also been graphed in Figure 70. The question now becomes: How have we changed the graph for $y = x^2$ to get the graph for $y = \sqrt{x}$? Notice the straight diagonal line passing through the origin. This line is our identity line which we mentioned previously. Because every x value is mapped onto the identical y value, the line is represented by the function $y = x$. If we take the graph for $y = x^2$ and reflect it around the identity line we get the new function $y = \sqrt{x}$. We call this new function, $y = \sqrt{x}$, the inverse function of $y = x^2$. Conversely, the function $y = x^2$ is the inverse function of $y = \sqrt{x}$. Reflecting a function around the identity line amounts to redrawing the graph with the x and y values swapped or exchanged. The inverse function has the same shape as the original function, but is reflected to a new orientation on the graph.

Inverse functions, as reflections of each other around the line $y = x$, answer opposite questions for us. We can now appreciate why we must have a one-to-one function to derive its inverse function. Consider the parabola $y = x^2$ when we allow x to be negative. This function is graphed in Figure 71. We have also shown the reflection of $y = x^2$, which gives us its inverse. This inverse is the parabola on its side in Figure 71. But this particular parabola is not a function because it fails the vertical line test. Hence, if we try to take the inverse of a function which is many-to-one, we end up with a relationship that is one-to-many and violates the definition of a true function. In other words, we end up with a relationship such that at least one x value is associated with two or more y values. Therefore, all inverse functions are produced from functions that are one-to-one, and every one-to-one function has its own inverse function.

The function $y = x^2$ tells us how to get y when we already know x. It says to multiply x by itself. For example, if we know that x is 3.4, then we know that y must be 3.4^2, or 11.56. What if we know y but not x? This is a more difficult problem. In our example we would have $11.56 = x^2$, and we are expected to find x. But, if we use the inverse function, the job becomes easy. The inverse function is $y = \sqrt{x}$, and since we swapped x and y to get it, we must now swap the x and y in the example. Hence, we have $y =$

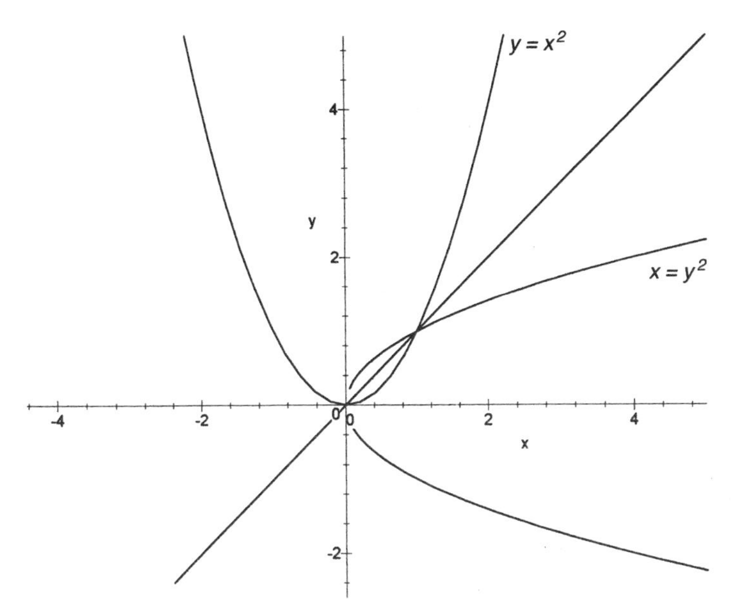

FIGURE 71. The reflection of the graph of $y = x^2$ for all values of x around the identity line to generate $x = y^2$. This demonstrates that a function that is not one-to-one, when reflected around the identity line, generates a graph that is not a true function.

$\sqrt{11.56}$. I know how to solve this problem; I extract the positive square root of 11.56, which is 3.4.

Therefore, the inverse function answers the opposite question of its original function. Many times we have a function and know what y is, but don't know x. We can solve such a problem if we can find the inverse function, which allows us to directly solve for the unknown. If $f(x)$ denotes a function then $f^{-1}(x)$ denotes its inverse.

LOGARITHMS

Now for a very surprising and pleasing result. Remember those awful things called logarithms that you were expected to learn, and you just knew they were too complex for any normal human to understand? We are about to make mincemeat of them. What is the inverse function for an exponential function? It is just a logarithmic function! This means that we can take the graph of the exponential function and reflect it around the

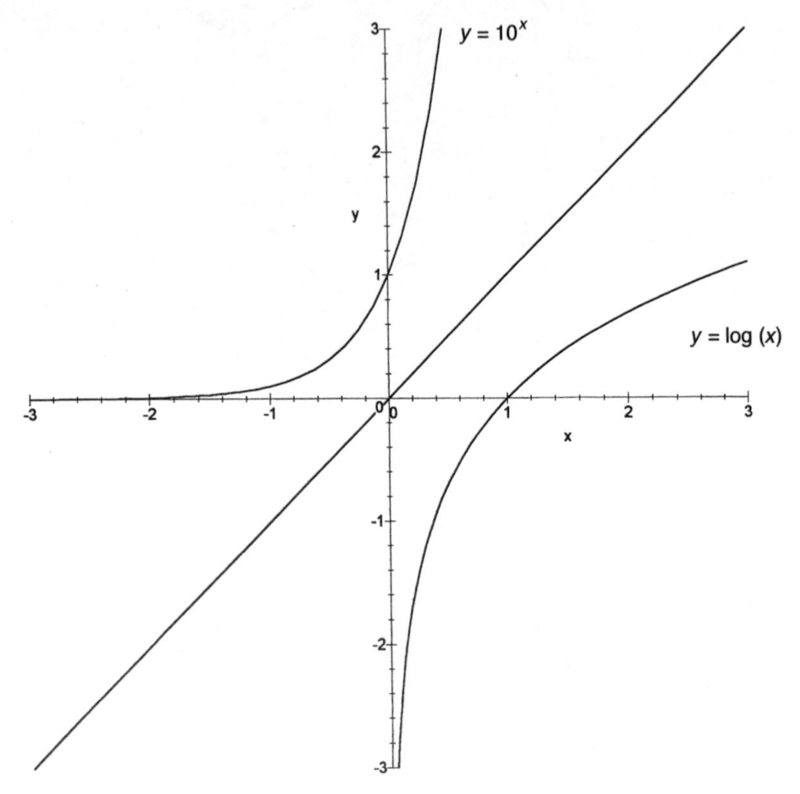

FIGURE 72. The graph of the exponential function $y = 10^x$ is reflected around the identity line to become the logarithmic function, $y = \log_{10}(x)$.

identity line, giving a new graph with the same general shape but a new orientation. This graph becomes our logarithmic function (Figure 72). So, all our work with functions is now paying off. Consider the exponential function, $y = 10^x$. If someone said that x was 1.2, I could figure out what y was. I can change the decimal 1.2 to the fraction $6/5$. I then raise 10 to the sixth power and extract the fifth root. This yields the number 15.8489. . . . Of course, those using a scientific calculator can just plug in $10^{1.2}$ to get the answer.

What if someone gave us the problem $32.991 = 10^x$? Here we have the y to an exponential function but not the x. Finding the x in this problem is more difficult. One indirect method is to experiment by trying different values for x to see how close they are to 32.991. Yet, this is time consuming, and a very hit-or-miss approach. But, if we define the inverse function to $y = 10^x$ we can solve the problem directly. That is just what a logarithm is—the inverse of an exponential function. We write the logarithmic function that is the inverse to $y = b^x$ as $y = \log_b(x)$, which means that the base b, raised to the y power, is equal to x. Hence, the two following expressions contain the same information, and each implies the other:

$$y = \log_b(x) \quad \text{implies} \quad x = b^y$$

From this we see that *logarithms are really exponents in disguise*. Now we use the inverse function of $y = 10^x$, namely $y = \log_{10}(x)$, to solve our problem. We know x is 32.991. Therefore, we solve $y = \log_{10}(32.991)$. Plugging this into my scientific calculator I get $y = 1.518395\ldots$. This tells me that $32.991 = 10^{1.518395\ldots}$. We call the logarithmic function with a base of 10 the common logarithm, and we generally show it as just $y = \log x$ without showing the base 10 as a subscript. Therefore, whenever you run across the logarithmic symbol without a base, assume it is a base of 10.

In the above example we used an exponential function with base of 10. We can do the same with The Exponential Function, using a base of e. Figure 73 shows The Exponential Function and its inverse function, the natural logarithmic function. The natural logarithmic function has the symbol $y = \ln(x)$. These two functions, each the inverse of the other, may well be the most important functions in mathematics—after the differential and integral functions of calculus, which we will be talking about soon. Many processes, both in the physical world, and in the world of pure mathematics, seem to obey one of these two functions. We defined the number e as a limit, and we can also define the natural logarithm of a number, x, as a limit.[3]

$$\lim_{n \to \infty} [n(\sqrt[n]{x} - 1)] = \ln x$$

Since we have an infinite series expression for e^x, we can use it to represent any exponential function, a^x, as an infinite series. We begin with the simple identity, $x \cdot \ln(a) = x \cdot \ln(a)$. Since $\ln(e) = 1$, we can rewrite the identity as $x \cdot \ln(a) = x \cdot \ln(a) \cdot \ln(e)$. Using the laws of logarithms we move the left hand terms up to exponents or $\ln(a^x) = \ln(e^{x \cdot \ln(a)})$. If two logarithms are equal, so are their arguments, so $a^x = e^{x \cdot \ln(a)}$. We now use the infinite series form for the Exponential Function to get:

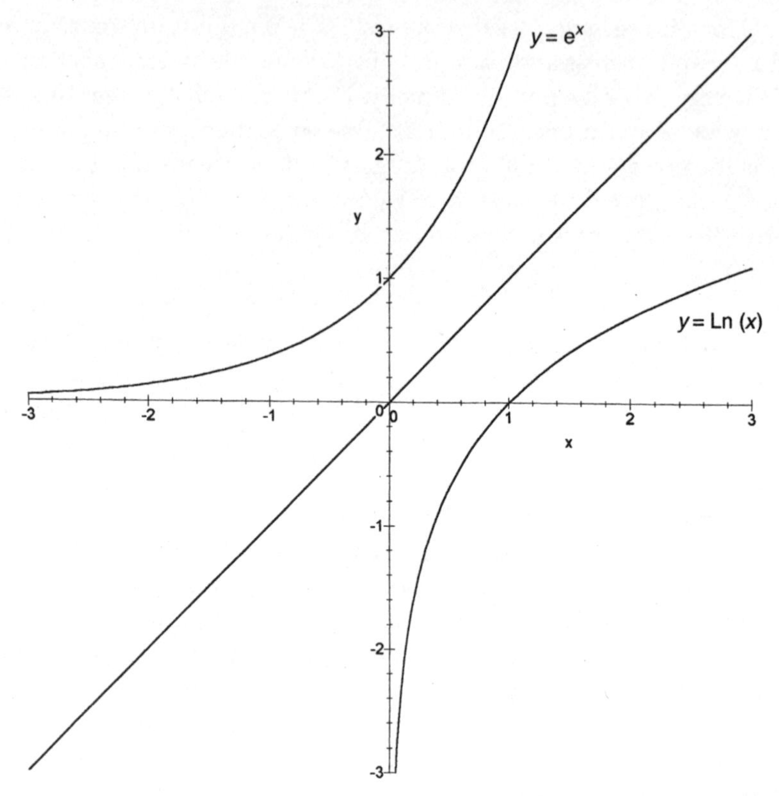

FIGURE 73. The graph of the exponential function $y = e^x$ is reflected around the identity line to become the logarithmic function, $y = \ln(x)$, one of the most important functions in mathematics.

$$a^x = 1 + x \cdot \ln(a) + \frac{[x \cdot \ln(a)]^2}{2!} + \frac{[x \cdot \ln(a)]^3}{3!} + \ldots$$

We can use the above infinite series to approximate any exponential function we want if we have a table or calculator that gives us the natural logarithm of numbers.

John Napier (1550–1617) was a Scotsman and amateur mathematician who published the first work on logarithms in 1614. The immediate power of using logarithms was to simplify multiplication, for logarithms change multiplication problems into addition problems. Their introduction by

Napier was well timed, for Johannes Kepler (1571–1630) was struggling to compute the orbit of Mars around the Sun. This involved long and distasteful multiplications, but using logarithms Kepler was able to complete the job and develop his three laws of planetary motion.

For an example of the importance of the natural logarithmic function, we can look to pure mathematics and the field of number theory. The deepest mystery within number theory is the distribution of the prime numbers. We have already talked about prime numbers and their importance in number theory since all positive integers factor into a unique set of prime numbers. Euclid proved over two thousand years ago that an infinite number of primes exist, but just where these primes occur within the number sequence is still a mystery since they have no predictable pattern. A function that measures the density of prime numbers within the number sequence is called the Prime Counting Function, which has the symbol of $\pi(n)$. This is not to be confused with the ratio π, which also uses the same Greek letter, but alone. The Prime Counting Function, $\pi(n)$, is defined as the number of primes less than n. Hence, $\pi(10) = 4$, since there are four primes less than or equal to 10, namely 2, 3, 5, and 7. For $n = 100$ we have $\pi(100) = 25$, which says there are 25 primes less than 100.

No one knows how to compute the Prime Counting Function exactly for large n without actually examining all the numbers less than n and counting those that are prime. There is no formula known that will do this in a fast, convenient manner. However, we do have a famous theorem discovered by the great mathematician Carl Gauss (1777–1855).

> *Prime Number Theorem*: The number of primes less than n is approximately n divided by the natural logarithm of n or
>
> $$\pi(n) \approx \frac{n}{\ln(n)}$$

As the natural number sequence increases, the number of primes less than n increases approximately as the ratio of n divided by the natural logarithm of n. This approximation is very close when n is very large. This is an amazing theorem because it says that the density of primes within the natural numbers can be described by the natural logarithmic function, which in turn has a base equal to our fabulous number e. Here we see that e is not only related to growth and decay process in nature, but is intimately connected to the distribution of primes within the number sequence.

We have expressed both π and e in terms of an infinite series. Can we do the same for natural logarithms? Of course! Consider the following fine infinite series:

$$\sum_{n=1}^{\infty} \frac{(-1)^{n+1}}{n} = \frac{1}{1} - \frac{1}{2} + \frac{1}{3} - \frac{1}{4} + \frac{1}{5} - \frac{1}{6} + \ldots = \ln 2$$

This is very strange, indeed. The harmonic series diverges, but the alternating harmonic series converges to the natural logarithm of 2. Another magnificent infinite series involving the natural logarithm is:

$$\sum_{n=1}^{\infty} \frac{(-1)^{n+1}(n!)^2}{n^2(2n)!} = \frac{1}{2} - \frac{1}{24} + \frac{1}{180} - \frac{1}{1120} + \frac{1}{6300} - \ldots = 2 \cdot (\ln \phi)^2$$

where ϕ is the Golden Mean. But how can this be? Not only do we have an infinite series connected to a natural logarithm, but also connected to our astounding Golden Mean, which we know is connected to so much else!

We will mention here one more application of the number e. The formula for compound interest is:

$$A = P(1 + R/n)^{nt}$$

where A represents the amortized amount, or that amount we either owe or collect at the term of the loan, P is the principal, R is the yearly interest rate, n is the number of compounding periods per year, and t is the number of years. For example, a \$3000 loan for five years ($t = 5$) at 12 percent yearly interest ($R = .12$) and compounded monthly ($n = 12$) would be:

$$A = 3{,}000(1 + .12/12)^{12 \cdot 5} = \$5{,}450.09$$

This formula looks suspiciously like the limit definition for e. If we compute interest that is compounded continuously—that is, compounded for every single instant of time over a specific period—then the compounding formula becomes: $A = Pe^{rt}$, again where P is the principal, r is the interest rate, and t is the number of years. Thus, to compound our loan with a principal of \$3,000 continuously for five years at a 12 percent interest rate would yield an amortized amount of:

$$A = (3{,}000)e^{(.12)(5)} = \$5{,}466.35$$

Amazingly, continuous compounding of interest does not increase the amortized amount significantly beyond monthly compounding.

Throughout the remainder of the book we will point out interesting connections with the wondrous number *e*, for if we collected all the material on *e* together, it would be enough for an entire book.[4]

TRIGONOMETRY

Trigonometry is the study of the relationship between angles and lengths. Its roots go back to the ancient Babylonians, and perhaps further. As mentioned before, one of the most notable changes observed by our ancestors, who could actually see the stars at night (as opposed to us modern city dwellers who see mostly neon lights), was the progression across the sky of the moon and stars from one night to the next. In addition, ancients used the height of the sun above the horizon as the first clock to determine the time of day. The ancient Egyptians noticed that the Nile flooded when the star Sirius began rising in the early evening sky. Thus it became important for them to predict when this would occur by keeping records of the location of Sirius with respect to the sun. The first problem was to measure how far above the horizon a star, the moon, or the sun was at any particular time. From this developed the idea of an angle. The moon has a period of almost thirty days, hence this unit of time was used by many early societies as a basis for their first calendars. The Babylonians used twelve months of thirty days each, which yields a total of 360 days. Since this is short of a true year by just over five days, they were required to throw in an extra month every six years. The ancient Mayans used one calendar of eighteen twenty-day months, which also gives 360 days. However, they corrected their calendar by adding five feast days at the end of each year.

Probably because they had 360 days in a calendar year, the Babylonians divided the circle into 360 parts, which is the origin of our 360 degrees. This was a convenient unit of measure since the heavenly bodies move a distance of slightly less than one degree (360/365th) from one night to the next. A crude method to determine how high some sky object is above the horizon is to hold your thumb at arm's length and count how many thumb widths measures the distance. While this method is certainly inaccurate, it is surprising that for many individuals the distance of sky covered is frequently between one and two degrees per thumb width.

However, a more accurate method is needed, one that will not be subject to errors caused by using different measuring devices (or the thumbs of different people). If we draw a line from an observer to the

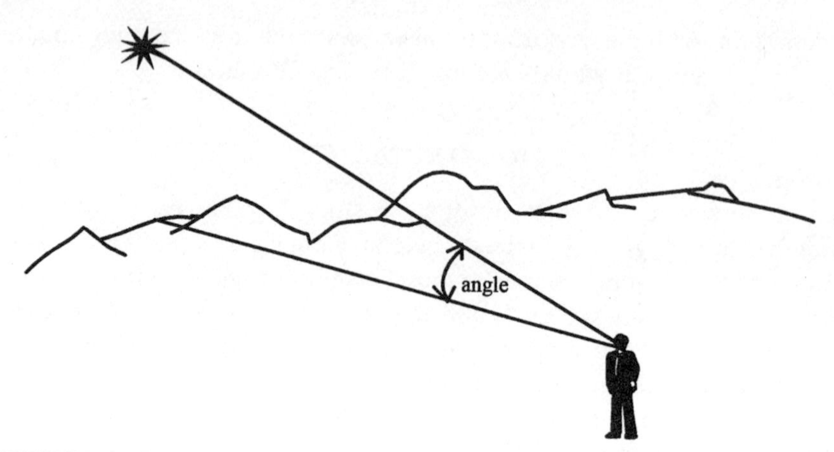

FIGURE 74. The angle trapped between a line-of-sight to the horizon and to a celestial object can be used to track the positions of heavenly bodies.

horizon, and a second line from the observer to the object we wish to measure (Figure 74), then, at the observer, the two lines trap an angle. The size of this angle is independent of the length of the arms of any instrument actually used for taking the measurement. Hence, this trapped angle is perfect for describing the positions of heavenly bodies.

While neither the Egyptians nor Babylonians defined the angle in our modern sense, they both tackled problems which involved right triangles and the ratio of their sides corresponding to different angles. Evidence of such work among the Babylonians can be found in the ancient clay tablet previously mentioned, the Plimpton Tablet No. 322, from the period between 1900–1600 B.C. Some analysts have claimed that the numbers in the columns correspond to the various dimensions of a right triangle, with each row corresponding to a right triangle containing an acute angle of between 45 degrees to 31 degrees.[5] This suggests the Babylonians were studying right triangles whose acute angle varied by a degree. Others dispute this interpretation of Tablet No. 322.[6]

The Egyptians, too, struggled with ratios of sides of right triangles. Problems in the Rhind Papyrus, dated around 1650 B.C., mention the concept of "seked," which is similar to the modern concept of the ratio of "rise" to "run," or slope. The Egyptian "seked" was the reciprocal of our slope, and was used to insure uniform slope in the construction of pyramids. For example, Problem 56 reads, "Example of reckoning a pyramid. Height 250, base 360 cubits. What is its seked?"[7]

The first serious study of angles determined the length of a chord cut off by a central angle of a circle. The first recorded table of chords was that of Hipparchus about 140 B.C. However, the length of a chord for any given angle was measured in 60 units, for the radius of the circle was considered to be this size. This is clearly an influence from the Babylonians. Using this measure of 60 units for the radius, an angle of 1 degree produces a chord of length 1.04 units. It would appear that the choice of 60 for the radius was not arbitrary, but based on the crude assessment that the ratio of the circumference of the circle to its diameter was three. Ptolemy also produced a table showing the length of chords for various angles.

Modern trigonometry uses the half chord rather than the entire chord for its basic definition. In Figure 75 we have divided the angle in half and produced two right triangles. The angle α now corresponds to the length of the half chord divided by the length of the radius. We define this ratio as the "sine of α," which becomes our basic trigonometric function.

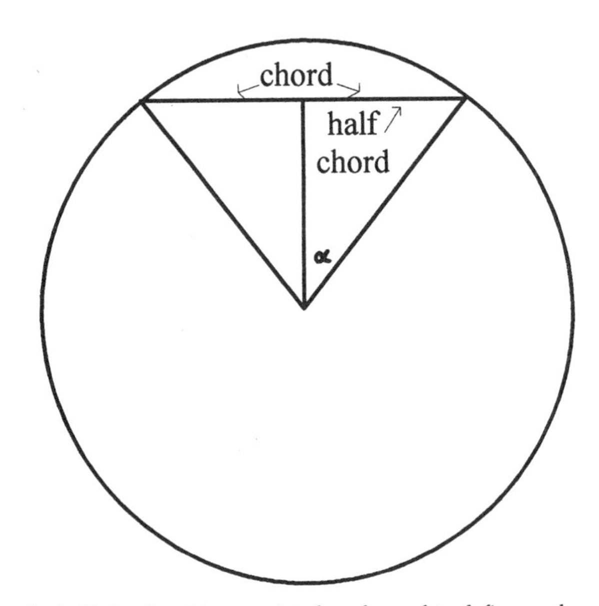

FIGURE 75. The half-chord and its associated angle used to define modern trigonometric functions.

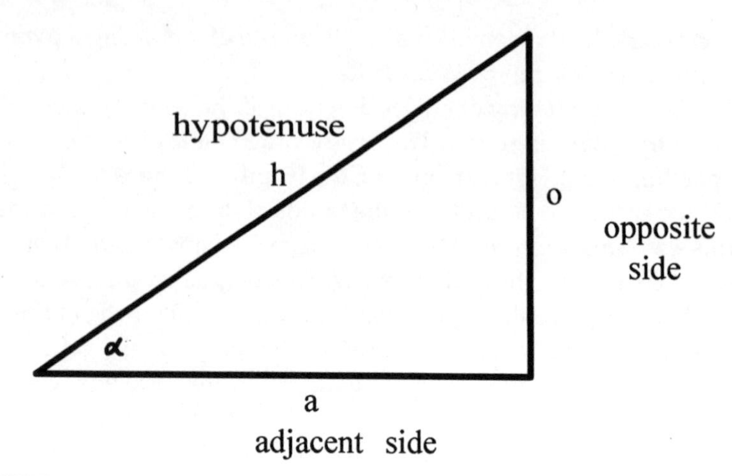

FIGURE 76. The right triangle used to define the basic trigonometric functions; sin (α) = o/h, cos (α) = a/h, and tan (α) = o/a.

Figure 76 shows a right triangle with an obtuse angle, α. The three sides are named the hypotenuse (h), the opposite side (o), and the adjacent side (a). The three basic trigonometric functions are defined as the ratio of the various sides of this triangle for any given angle, α. These functions are:

Name	Symbol	Ratio
sine .	sin (α)	o/h
cosine	cos (α)	a/h
tangent	tan (α)	o/a

These three functions not only play an important role in many areas of higher mathematics, they also have beautiful curves and numerous practical applications. Let's consider the sine function in more detail. In Figure 77 we have a circle whose center is the origin and whose radius is 1. This is called the unit circle. We can define the radius as the hypotenuse of our right triangle, the length along the x-axis as the adjacent side, and the length along the y-axis as the opposite side. The angle α is the angle of the right triangle located at the origin. Now when the hypotenuse is collapsed onto the x-axis, corresponding to an α of 0, then the length of the opposite side is also 0. Hence, sin (0) = 0, which simply says that the sine of 0 degrees is 0.

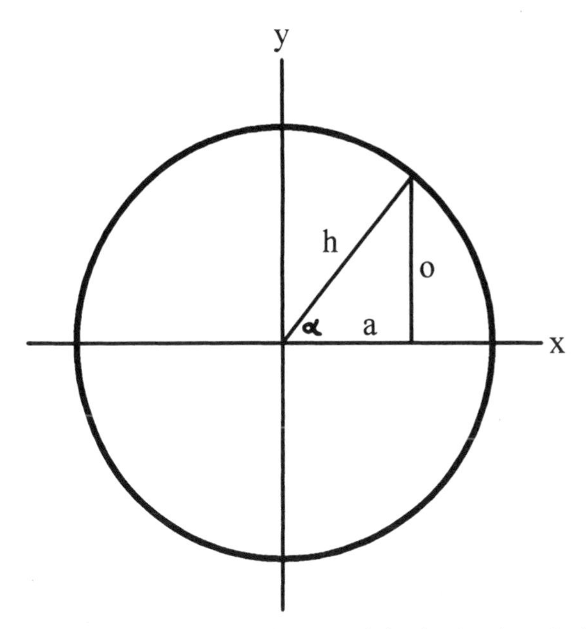

FIGURE 77. The trigonometric functions defined using the unit circle.

As we rotate the hypotenuse in a counterclockwise direction, allowing the angle α to grow in size, we can see that the opposite side will increase in length compared to the length of the hypotenuse, which stays constant. In Figure 78 we have plotted the magnitude of this growth on a new coordinate system where the value of α is plotted on the x-axis and the value of $\sin(\alpha)$ is plotted on the y-axis. As α rotates from 0 degrees through 90 degrees, the value of $\sin(\alpha)$ grows from 0 to 1. This growth is plotted on the right side of Figure 78.

Figure 79 shows how the magnitude of $\sin(\alpha)$ changes as we rotate α from 90 degrees to 270 degrees. Notice how the value of $\sin(\alpha)$ gently drops from 1 to 0 (at 180 degrees) and then continues down to -1. The sine of an angle is therefore a real number ranging from $+1$ to -1 in value, depending upon the magnitude of the angle. The sine is like an ocean wave that moves from a crest of $+1$ to a trough of -1, and back again. In Figure 80 we show the last 90 degrees of our rotation and how $\sin(\alpha)$ rises up to a value of 0 again. If we were to continue the rotation of our hypotenuse through another 360 degrees, the values of $\sin(\alpha)$ would make

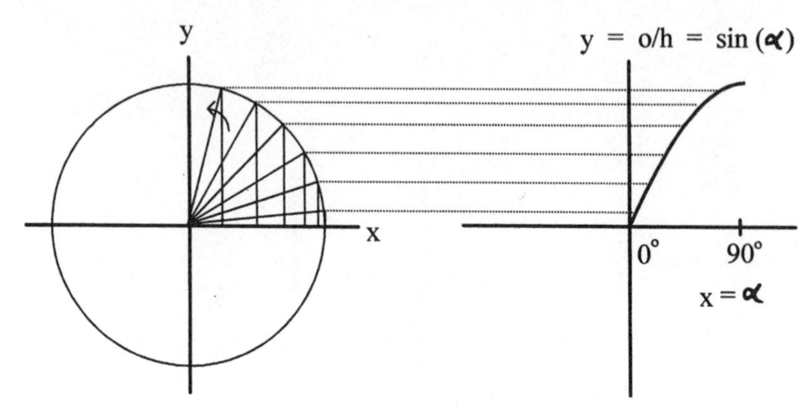

FIGURE 78. The generation of the sine curve. As the hypotenuse moves counterclockwise from the x-axis to a vertical position, the value of the sine function increases from zero to one. This corresponds to the value of α changing from zero to 90 degrees.

an exact replication. Hence, the sine function is a periodic function that will continuously repeat itself as we rotate the hypotenuse through ever larger angles. An example of the sine function from a minus 720 degrees to a positive 720 degrees (four full rotations of the unit circle) is shown in Figure 81. The sign function produces an especially lovely curve, one not duplicated by other elementary functions (with the exception of the cosine

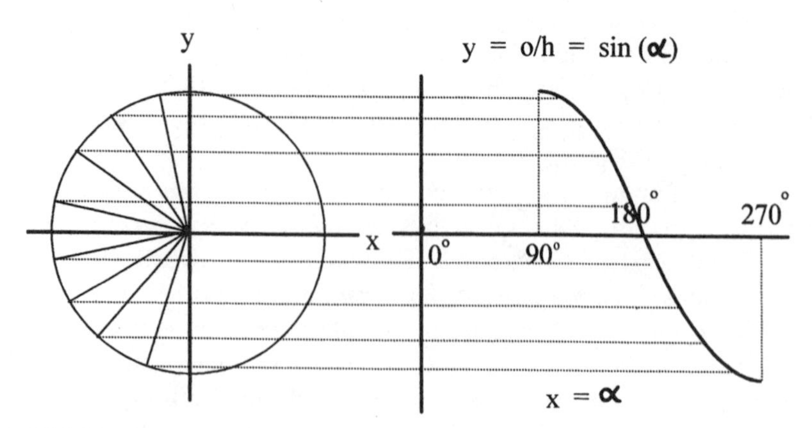

FIGURE 79. Generating the sine curve as α changes from 90 degrees to 270 degrees, corresponding to a change in $\sin(\alpha)$ from 1 through zero to -1.

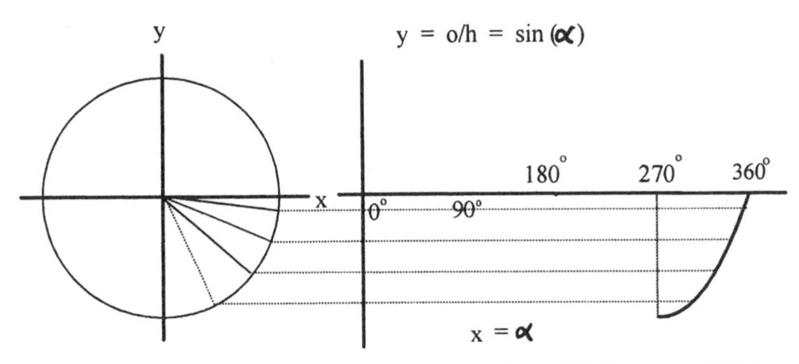

FIGURE 80. Generating the sine curve as α changes from 270 degrees to 360 degrees, corresponding to a change in sin (α) from −1 back to 0.

function). The sine function, and its siblings, the cosine and tangent, are extremely useful because they can be made to replicate many natural phenomena—from the action of ocean waves to the transmission of light. Such phenomena include the vibration of a drumhead, the changing of tides, and the fluctuation of voltage within your household current.

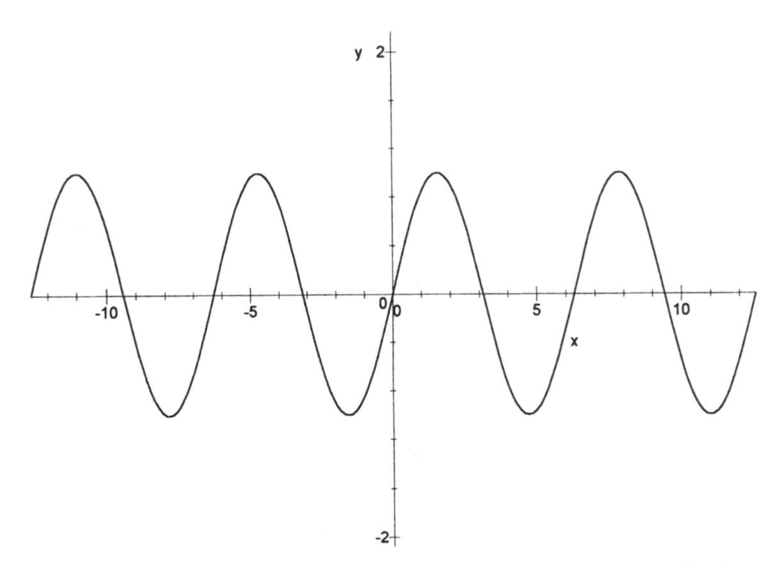

FIGURE 81. The graph of the sine curve as α changes from −720 degrees to +720 degrees, or four full rotations of the unit circle.

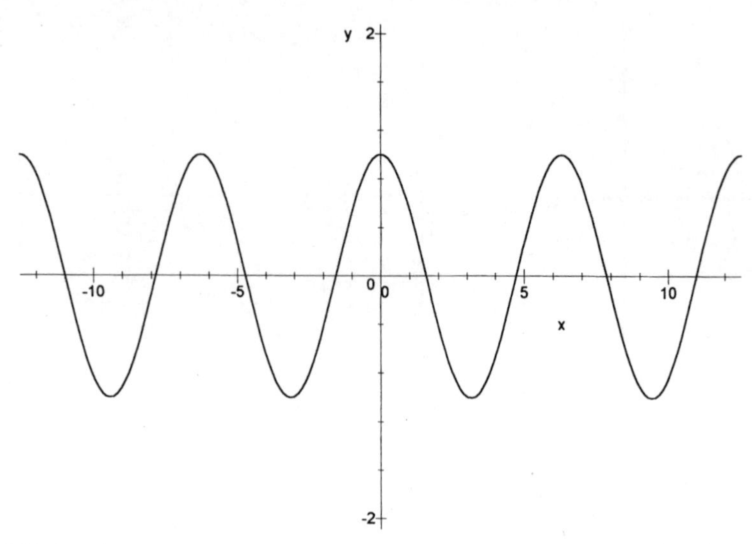

FIGURE 82. The graph of the cosine curve as α changes from -720 degrees to $+720$ degrees, or four full rotations of the unit circle. The cosine curve is identical to the sine curve but shifted toward the right by 90 degrees.

The graph of the cosine function is identical to that of the sign function but shifted toward the right by 90 degrees as in Figure 82. In many applications we don't express the angle α in degrees but in radians. An angle of one radian corresponds to an arc length equal to the radius of the circle, as in Figure 83. Converting 360 degrees (one complete rotation) to radians yields 2π radians (approximately 6.283 radians).

Examples of the use of trigonometric functions to model real world situations abound, and we will consider only a few. Trigonometry has been used for centuries in navigation. Suppose we are sailing in the ocean, but are uncertain if we are far enough offshore to avoid a dangerous coral reef located 1 mile from the beach. We know the cliffs at the water's edge are 400 feet high (Figure 84). We take a sighting from our ship and determine that the angle between the top and bottom of the cliffs is just 3 degrees. Since the tangent of an angle is the opposite side divided by the adjacent side, we know that $\tan 3° = (\text{height of cliffs})/(\text{distance to shore})$. Solving for the distance to shore we get: $D = 400'/\tan 3°$. We now look up the tangent of 3 degrees and divide that into 400 feet to get $D = 7{,}632$ feet. Hence, we know we are almost a half mile beyond the dangerous reef.

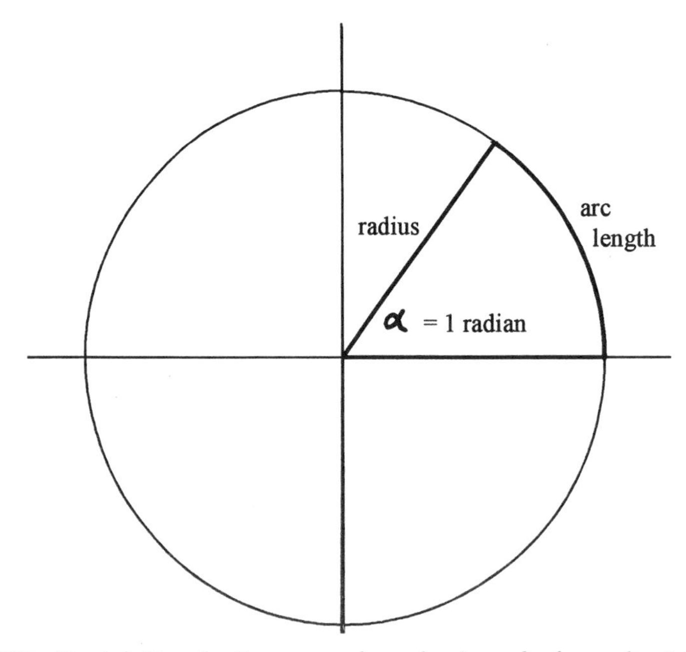

FIGURE 83. The definition of radian measure for angles. An angle of one radian is the angle trapped by the x-axis and the radius rotated through an arc length equal to the radius. One complete rotation (360 degrees) is equal to 2π radians.

The above example was a static model using a trigonometric function, but many nice dynamic models are available. We can model the motion of a weight suspended from a spring with either the sine or cosine function.[8] The vertical displacement of a weight bouncing on a spring is $y = A \cdot \sin(\omega t)$, where A is the maximum displacement of the weight, ω is the *angular frequency*, and t is the time. If we were to graph the value of y, the weight's vertical displacement, with time (t) as the x-axis, we get a curve very similar to that shown in Figure 81.

Our next example comes from the field of magnetism. The equation for the force acting on a charged particle, such as an electron, moving through a magnetic field is given by:

$$F = Bqv \cdot \sin\theta$$

where B is the magnetic induction (strength) of the magnetic field, q is the strength of the charge on the particle, v is the particle's velocity, and θ is

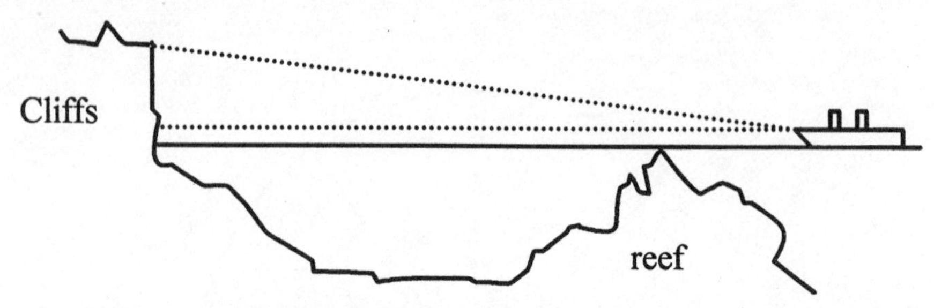

FIGURE 84. Using trigonometry to solve simple problems. The captain of the boat measures the distance to the shore by knowing the height of the cliffs and the angle trapped between the line-of-sight to the beach and the top of the cliff.

the angle between the particle's path and the direction of the magnetic field (Figure 85).

Our last example also comes from the field of magnetism and electronics. Suppose we have an electrical circuit that contains both a capacitor and an inductor as in Figure 86. If we close the switch on such a circuit, the electrical charge in the capacitor will discharge and flow through the

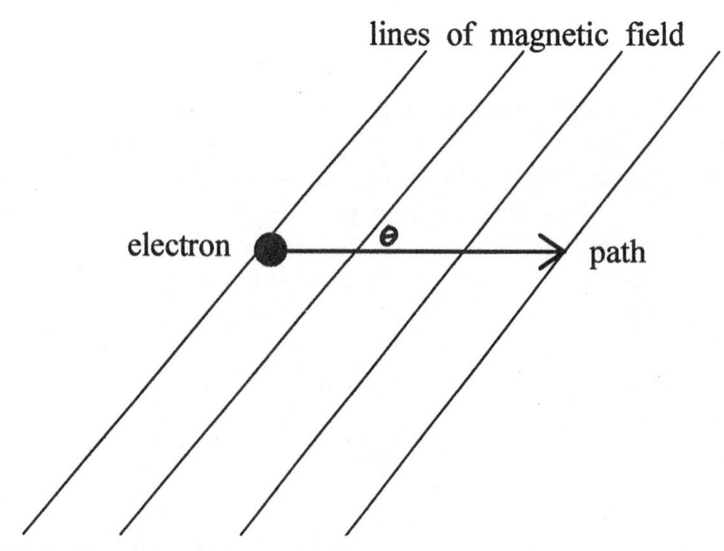

FIGURE 85. The angle made by an electron as it passes through a magnetic field. The force acting on the electron by the magnetic field is $F = Bqv \cdot \sin\theta$.

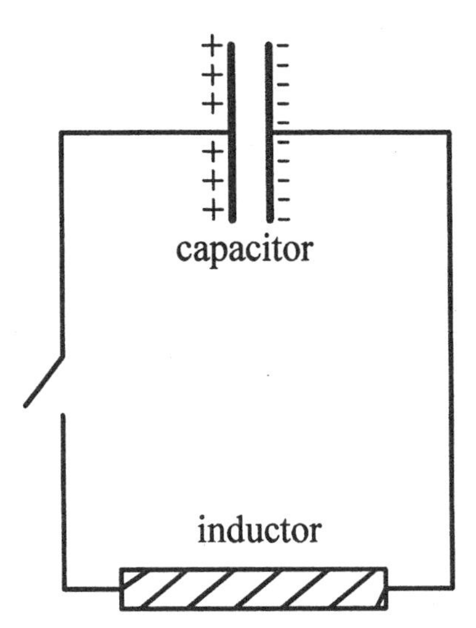

FIGURE 86. A circuit containing both a charged capacitor and an inductor. When the switch is closed, the capacitor discharges, sending a current through the circuit. When the current passes through the inductor, it creates a magnetic field which induces another current in the opposite direction, charging the capacitor. The capacitor discharges and the process repeats itself.

inductor. The flow of electricity through the inductor will cause the inductor to establish a magnetic field, which in turn induces another current into the circuit, recharging the capacitor with opposite polarity. The capacitor then discharges again, but in the reverse direction, and the whole process is repeated. Assuming there is no loss of energy, the flow of current in the circuit and the charge on the capacitor would oscillate indefinitely. The model we use to describe this oscillation is identical to the weight bouncing on a spring. The charge on the capacitor at time t is q and its equation is: $q = Q \cdot \sin(\omega t)$, where Q is the maximum charge on the capacitor, ω is the natural frequency of the circuit, and t is the time since the circuit was closed. When we consider the loss of heat due to resistance, each successive current flow in the circuit is weaker. The curve describing this current is a diminishing cosine curve (Figure 87).

This remarkable parallelism between a mechanical system of a weight bouncing on a spring and the electrical system of an oscillating current in a circuit can be found between many other physical systems. Such parallel-

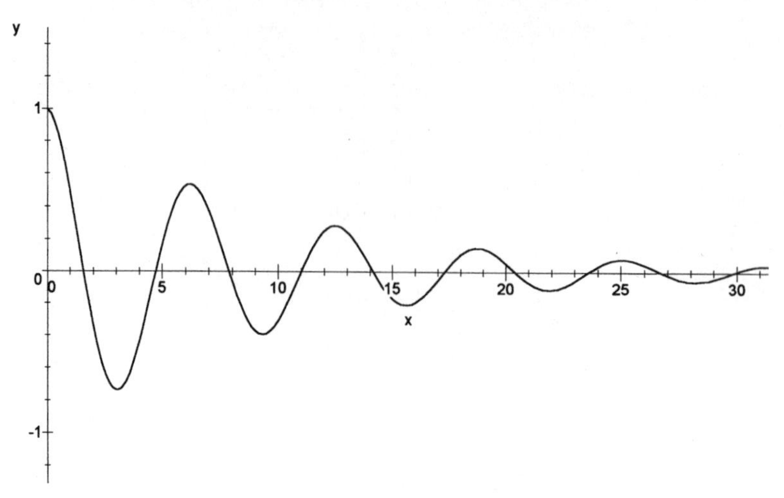

FIGURE 87. The dampening cosine curve used to model the circuit in Figure 86 when heat loss due to resistance causes the current to decrease after each cycle.

ism makes it possible to solve many difficult problems in physics by constructing the appropriate electrical circuit that mimics those problems, saving enormous time and resources.

We have been able to represent functions up to this point with infinite series, and of course we want to know if we can represent the trigonometric functions so. Two beautiful infinite series are defined for $\sin x$ and $\cos x$.

$$\sin x = x - \frac{x^3}{3!} + \frac{x^5}{5!} - \frac{x^7}{7!} + \ldots$$

$$\cos x = 1 - \frac{x^2}{2!} + \frac{x^4}{4!} - \frac{x^6}{6!} + \ldots$$

Exponential, logarithmic, and trigonometric functions are called *transcendental functions* as opposed to the *algebraic functions* which are created from the polynomials. The values we get from substituting specific values of x into transcendental functions are generally transcendental numbers. Setting algebraic functions equal to zero and solving for the unknown yields solutions that are algebraic numbers. Setting the transcendental functions equal to zero and solving yields transcendental numbers except in special cases. The algebraic functions together with the transcendental functions, including their inverses, constitute the elementary functions.

COMPLEX NUMBERS

To round out our discussion of the elementary functions we should make mention of a kind of number not yet considered. We have talked extensively about the number line and the real numbers that fill it up. For most purposes, the real number line serves our ends and allows us to carry out the computations we require. However, early mathematicians were perplexed and frustrated because of a simple problem in algebra.

Recapping our generation of numbers, we began as hunter–gatherers with the counting numbers. When they were insufficient to account for all our needs we invented positive fractions. Then when these positive numbers were insufficient, we extended the number system to include the negative numbers. The Pythagoreans discovered another type of number, the irrational number, and we added them to our system. Now we must consider a problem which the real numbers can't touch.

At first glance the equation $x^2 + 1 = 0$ appears to be an exceedingly easy equation to solve for x. We just subtract 1 from both sides to yield $x^2 = -1$, then we take the square root of both sides to get $x = \pm\sqrt{-1}$.

Here is where we run into problems. The square root of negative one is a number, which when multiplied by itself, will yield a negative one. But wait, any time I multiply a real number by itself, whether negative or positive, I get a positive result. Hence, by multiplying the real numbers by themselves, I'm guaranteed never to get a negative number. How are we to solve this simple equation? The answer is to invent more numbers.

Carl Friedrich Gauss, noted earlier, was a child prodigy born to peasant parents. Through the help of his mentor, the Duke of Brunswick, he was able to receive a good education at the University of Göttingen, Germany. Considered to be one of the greatest mathematical minds to have ever lived, Gauss contributed significantly to a number of different fields in both mathematics and physics. To solve our little conundrum, he simply took two axes and crossed them at right angles. Thus he used a device nearly identical to the coordinate system first suggested by Descartes and Fermat, yet he defined the axes to provide an entirely new set of numbers. He defined the x-axis as the real number line, a line we are well familiar with. The vertical or y-axis he defined as the imaginary axis (a poor choice of words).

Figure 88 shows such a coordinate system. Gauss then went on to define one unit of length along the vertical axis as $\sqrt{-1}$, which had already been given the designation of i by Euler in 1777.[9] Multiplying a number by i ($\sqrt{-1}$) caused a counter-clockwise rotation in the plane of 90 degrees.

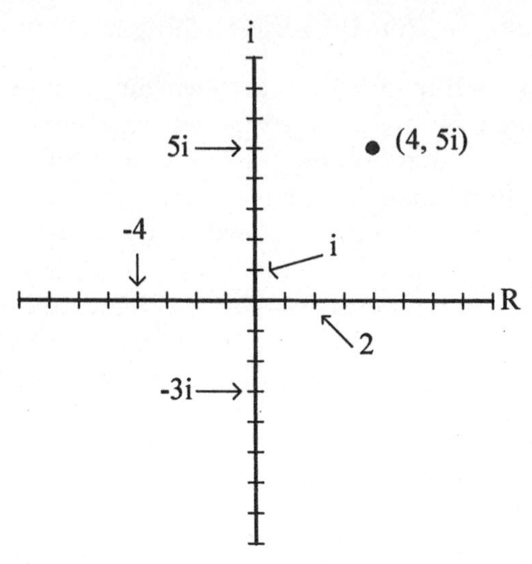

FIGURE 88. The Gaussian or complex coordinate system. Each complex number in the plane is uniquely defined by an ordered pair of real numbers, written as either (a,b) or $a + bi$, where a represents the real part of the complex number and b represents the imaginary part.

Thus we have $i \cdot 1 = i$ (Figure 89). When we multiply i by itself we rotate another 90 degrees to end up at -1, or $i \cdot i = i^2 = -1$. This definition gives the number i the very property we so eagerly need—that is, a number which when multiplied by itself is -1.

Using the idea of coordinates from analytic geometry, we define every point on the plane as a complex number made up of two components: the x-coordinate is the real component, while the y-coordinate is the imaginary component. Those numbers on the x-axis are complex numbers where the imaginary part is zero, i.e. real numbers. Those points on the y-axis are pure imaginary numbers with no real part. Every complex number in the complex number plane can be written as $a + bi$, where a is the real part and b is the imaginary part.

We have the nice identity $(a + bi) \cdot i = ai + bi^2 = -b + ai$. The effect of multiplying any complex number $a + bi$ by i yields a new complex number $-b + ai$, which is a rotation of the original number through 90 degrees.

Now that we have our complex numbers we can introduce one of Gauss' famous theorems:

> Every polynomial of degree n has n roots.

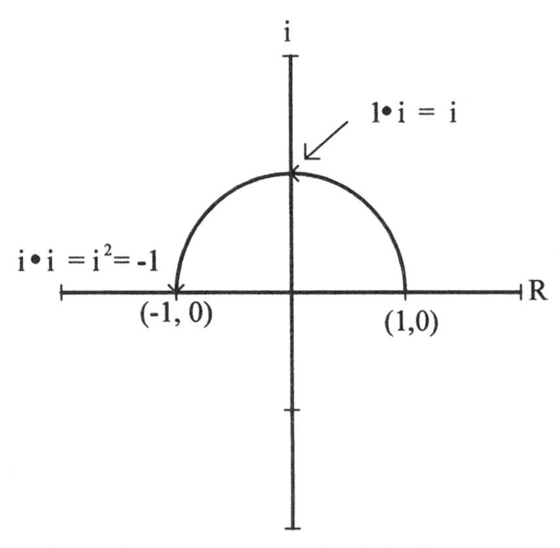

FIGURE 89. Multiplication by the imaginary i unit is defined as a counterclockwise rotation in the complex number plane. This causes i to have the desired characteristic of $i^2 = -1$.

Some or all of these roots can be complex numbers, e.g., lying off the real number line and on the complex plane. In turn this means that extracting the nth root of a real number produces n different complex numbers. This is because the equation $x^n - C = 0$ (which is a polynomial equation of degree n) is solved as $x = \sqrt[n]{C}$. Hence, the equation must have n roots, which in this simple equation all turn out to be equally spaced around the origin of the complex number plane. For example, $x^3 - 8 = 0$ is solved as $x = \sqrt[3]{8}$. We know that one root or solution is simply 2, but what are the others? The other two roots are the complex numbers $-1 + i\sqrt{3}$, and $-1 - i\sqrt{3}$, both of which have been graphed in Figure 90. These three roots are all an equal distance from the origin, a distance of two units, and they are evenly spaced about the origin. The nth root of a number, C, will produce n different roots, all of which will be positioned an absolute distance of $\sqrt[n]{C}$ from the origin and equally spaced about the origin.

Can we operate with complex numbers just as we operate with real numbers? Yes. We define for complex numbers all the customary operations: addition, subtraction, multiplication, division, exponentiation, and extracting roots. For example, to add two complex numbers, $a + bi$ and $c + di$, we simply add the corresponding real parts and the imaginary parts to get:

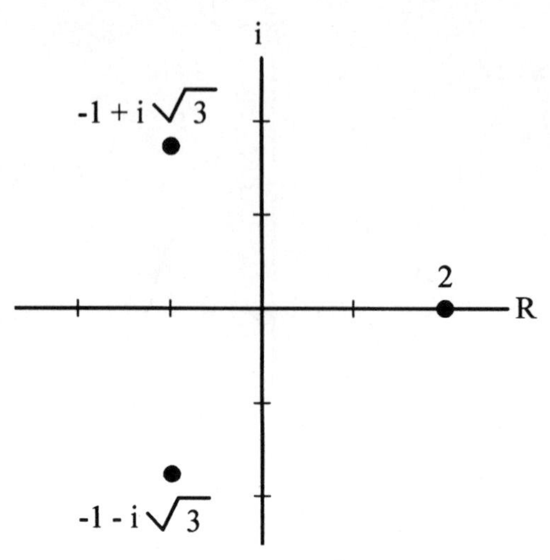

FIGURE 90. The three solutions to the equation $x^3 - 8 = 0$ plotted in the complex plane.

$$(a + bi) + (c + di) = (a + c) + (b + d)i$$

We define multiplication of two complex numbers in the following fashion:

$$(a + bi) \cdot (c + di) = (ac - bd) + (ad + cb)i$$

To see that the above identity holds, all we have to do is treat the two complex numbers like two binomials, multiply them out, convert the i^2 term to -1, and collect like terms. Division of complex numbers has a daunting formula that really isn't worth learning. If $a + bi$ is a complex number, then $a - bi$ is its conjugate. Hence, we find the conjugate of a complex number by reversing the sign in front of the imaginary term. To divide one complex number by another, $(a + bi)/(c + di)$, we simply multiply both the numerator and denominator by the conjugate of the denominator. This causes the denominator to become a real number without an imaginary part. Hence, we can divide the fraction into a real and imaginary part which is now our quotient.

$$\frac{(a + bi)}{(c + di)} = \frac{(a + bi)(c - di)}{(c + di)(c - di)} = \frac{ac - adi + bci - bdi^2}{c^2 + cdi - cdi - d^2i^2}$$

$$= \frac{(ab + bd) + (bc - ad)i}{c^2 + d^2} = \frac{(ac + bd)}{c^2 + d^2} + \frac{(bc - ad)i}{c^2 + d^2}$$

It is actually easier to divide one complex number by another than it is to demonstrate it!

Once we have defined the operation of multiplication we can define raising a complex number to a power (exponentiation) and extracting roots of a complex number. But now we reach an interesting question. If we can raise a complex number to a power that is represented by a real number, can we raise a real number to an exponent that is a complex number, and can we raise a complex number to an exponent that is, itself, a complex number? In other words, do the following expressions have any meaning?

$$5^{3+2i} \text{ and } (4 - 2i)^{1-6i}$$

Yes, such forms are defined! In fact, we have the very strange case where $i^i = e^{-\pi/2}$, which is approximately 0.207879 . . .—a real number! Raising real and complex numbers to complex powers is a fairly sophisticated business but can be great fun.

Now we have the opportunity to combine complex numbers with trigonometric functions and The Exponential Function to generate one of the most beautiful and elegant equations humankind has ever discovered. The complex number $x + yi$ is represented in the complex plane as a point which is x units along the x-axis and y units along the imaginary axis. In Figure 91 we have graphed our imaginary number in the complex plane and drawn a line from the origin to the point. This line is the hypotenuse to a right triangle with x and y as the lengths of the two legs. We label our hypotenuse r. The angle from the x-axis to the hypotenuse r is θ. We can represent the sine and cosine of θ as $\sin\theta = y/r$ and $\cos\theta = x/r$. Now solving for x and y we get: $x = r\cdot\cos\theta$ and $y = r\cdot\sin\theta$. Thus we have the identity:

$$x + yi = r\cdot\cos\theta + r\cdot i\cdot\sin\theta$$

Factoring out the r we get: $x + yi = r(\cos\theta + i\cdot\sin\theta)$, one of the most basic identities in mathematics, for it allows us to represent complex numbers using trigonometric functions. We are about to see how useful this is.

Now let's ask ourselves what the weird expression $e^{\pi i}$ could possibly be, for it combines our two most important constants, π and e, plus the imaginary unit, i. To figure out what this expression is, we are going to expand $e^{\pi i}$ using the infinite series for e^x.

$$e^{\pi i} = 1 + \pi i + \frac{(\pi i)^2}{2!} + \frac{(\pi i)^3}{3!} + \frac{(\pi i)^4}{4!} + \frac{(\pi i)^5}{5!} + \frac{(\pi i)^6}{6!} + \ldots$$

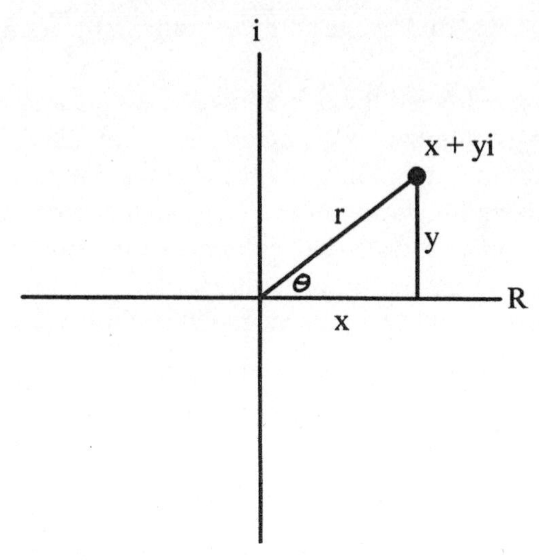

FIGURE 91. Defining a complex number, $x + yi$, in terms of trigonometric functions yields $x + yi = r \cdot \cos\theta + r \cdot i \cdot \sin\theta$.

When we get rid of the parentheses we have:

$$e^{\pi i} = 1 + \pi i + \frac{\pi^2 i^2}{2!} + \frac{\pi^3 i^3}{3!} + \frac{\pi^4 i^4}{4!} + \frac{\pi^5 i^5}{5!} + \frac{\pi^6 i^6}{6!} + \ldots$$

However, $i^2 = -1$, $i^3 = -i$, and $i^4 = 1$. In every fourth term in the above infinite series we can make the same substitution of one of these terms. This yields:

$$e^{\pi i} = 1 + \pi i - \frac{\pi^2}{2!} - \frac{\pi^3 i}{3!} + \frac{\pi^4}{4!} + \frac{\pi^5 i}{5!} - \frac{\pi^6 i}{6!} - \ldots$$

Now we simply group together all the terms without an i and all the terms with an i to get:

$$e^{\pi i} = \left(1 - \frac{\pi^2}{2!} + \frac{\pi^4}{4!} - \ldots\right) + \left(\pi i - \frac{\pi^3 i}{3!} + \frac{\pi^5 i}{5!} - \ldots\right)$$

We now factor out the i term from the right parentheses:

$$e^{\pi i} = \left(1 - \frac{\pi^2}{2!} + \frac{\pi^4}{4!} - \ldots\right) + i \cdot \left(\pi - \frac{\pi^3}{3!} + \frac{\pi^5}{5!} - \ldots\right)$$

At once we notice the most amazing thing! The expression inside the left pair of parentheses is just the expression for the infinite series for cosine, while that in the right pair of parenthesis is just the expression for the sine function. Hence we have: $e^{\pi i} = \cos(\pi) + i \cdot \sin(\pi)$. Therefore, we have expressed $e^{\pi i}$ in terms of sine and cosine. But we don't have to stop here, for $\cos(\pi) = -1$ and $\sin(\pi) = 0$. This means that $e^{\pi i} = -1$, or

$$e^{\pi i} + 1 = 0$$

Thus, in one identity we have combined the two most important constants in mathematics: e and π, the fabulous unit for imaginary numbers $\sqrt{-1}$, the fundamental operation of addition, the number 1, equality, and zero. Can it get any better than this? But we must give credit where it is due. It was our friend, the prolific Leonhard Euler who discovered this wonderful identity, an identity which has been thrilling mathematicians for hundreds of years.

7

Stretching Space

In previous chapters we introduced Fermat's and Descartes' splendid rectilinear coordinate system, which allows us to graph various symbolic relationships. However, we can't stop with just the simple Cartesian system, for the curious cannot leave well enough alone. We must see if it is possible to alter the rectilinear coordinate system into something even more interesting. For example, what happens when we use a y-axis that is not perpendicular to the x-axis? Figure 92 shows such a system where the y-axis intersects the x-axis at 60 degrees. In the second quadrant we have drawn a normal circle whose equation is:

$$(x + 4)^2 + (y - 4)^2 = 4$$

The oblong shape is achieved by using the above equation but with the new oblique y-axis. This new shape is that of an ellipse, still one of our conic sections. Remember that a circle is a special case of an ellipse where the plane intersecting our two cones is perpendicular to the axis of our cones. If we tilt this intersecting plane, then we generate an ellipse instead of a circle. Therefore, the effect of tilting our y-axis is the same as tilting our cutting plane, changing a circle into an ellipse. We can see that tilting our y-axis has had the effect of stretching the shape of the circle into something different.

However, we don't have to be content simply to rotate the y-axis, leaving it straight, for we can bend the y-axis into the shape of a parabola. In Figure 93 we have done just that. Notice that the circle is now stretched into a shape similar to that in Figure 92. Yet, in this case the circle has not been changed into a simple ellipse. Because we have used a curve (a parabola, for example) as the y-axis the circle has been deformed into a much more complex object. The equation needed to describe this shape in rectilinear coordinates is an equation of the fourth degree, whereas all the conics, including the ellipse, are equations of the second degree. Thus an

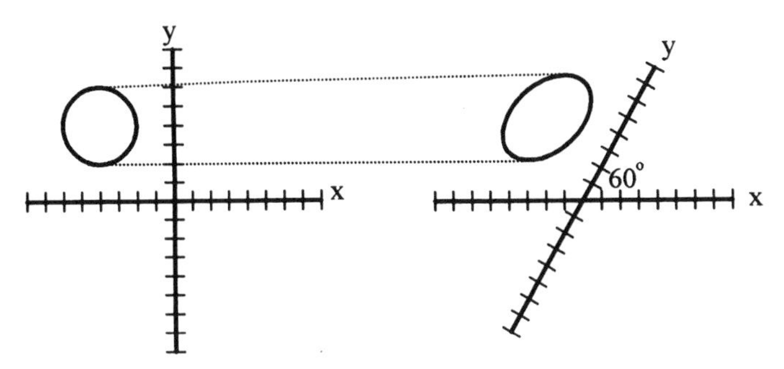

FIGURE 92. Tilting the y-axis in the Cartesian coordinate system from 90 degrees to 60 degrees causes the circle to become an ellipse.

equation in the second degree (circle) combined with the parabolic y-axis has resulted in a fourth degree curve.

We can get even fancier. In Figure 94 we have changed both the y-axis and the x-axis into parabolas. Now our poor circle has been stretched into a strange oblong blob. Again, this new curve will be of the fourth degree and

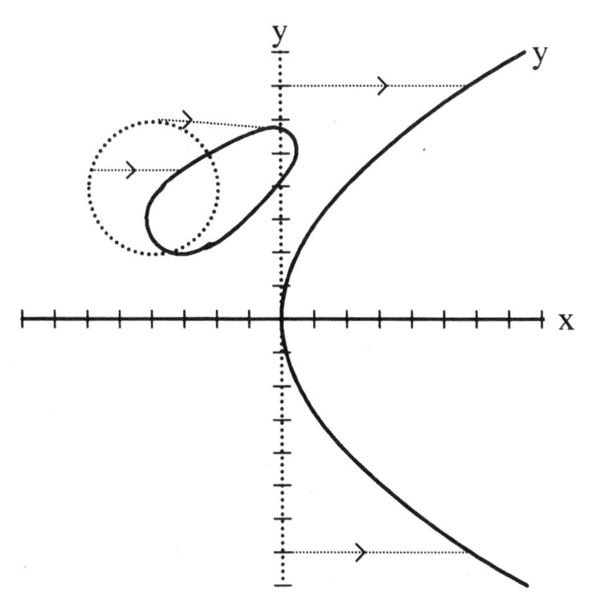

FIGURE 93. Using a y-axis in the shape of a parabola causes the circle to deform into an oblong which is represented by a fourth degree equation.

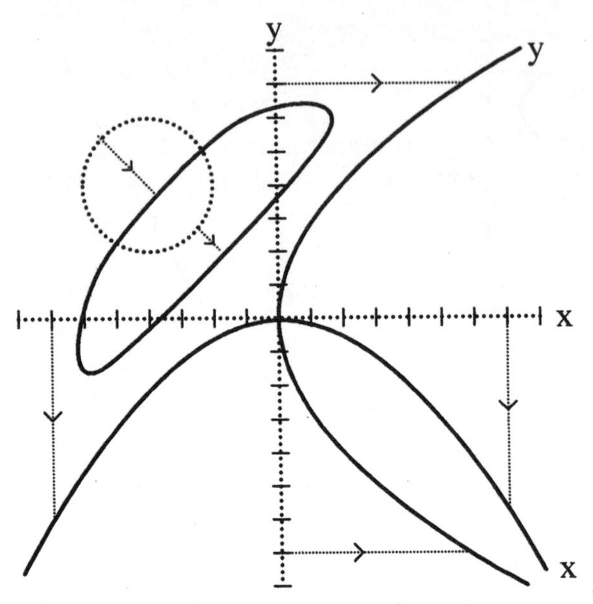

FIGURE 94. Changing both the x-axis and the y-axis into parabolas further distorts a circle.

not the second. The wildness of our graphing technique is only limited by our own imaginations.

Carl Gauss suggested using a more generalized coordinate system where the axes could be curves, such as parabolas. He invented differential geometry which describes all possible geometries of a surface. The key to any geometry is the rule or formula for finding the distance between two points. Gauss considered the distance between any two points that were infinitesimally close and designated this distance as ds, called the differential for s. In other words, he was considering the characteristic shape of a space on the microcosmic, rather than the macrocosmic scale, and was concerned with the basic curvature of the space more than its global shape. In Euclidean geometry, which deals with flat surfaces, the Pythagorean Theorem expresses this rule for distances between points: $ds^2 = dx^2 + dy^2$. Here s is the hypotenuse and x and y are the legs of a right triangle (Figure 95).

Gaussian differential geometry generalized this to be the rule: $ds^2 = g_{11}dx^2 + g_{12}dxdy + g_{21}dydx + g_{22}dy^2$. In this rule the x and y are not restricted to straight and perpendicular axes, but can be curves. The four g_{ij} in his

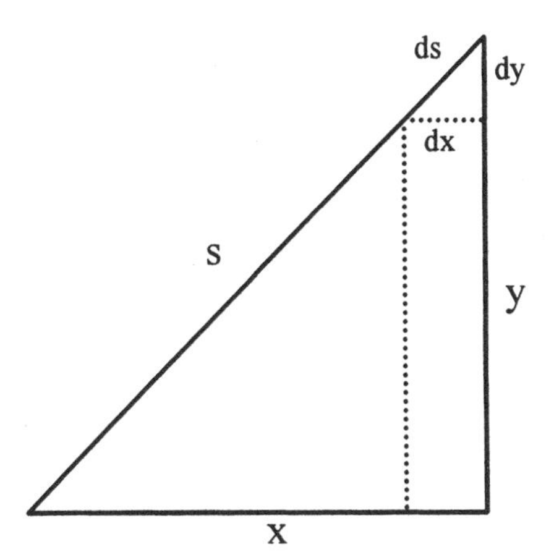

FIGURE 95. Gaussian differential geometry defines the characteristic shape of a space on the microcosmic scale. When the space is "flat" Euclidean space, then the distance between any two infinitesimally close points satisfies the Pythagorean theorem for infinitesimal distances or $(ds)^2 = (dx)^2 + (dy)^2$.

expression are either constants or functions dependent on x and y, and they completely determine the curvature of the surface of the space. From them Gauss was able to calculate the surface curvature. If this curvature is zero at every point, then the surface is Euclidean or "flat."

POLAR COORDINATES

We don't even have to stay with a coordinate system that is defined by two intersecting lines or curves. In 1671 Isaac Newton became one of the first to mention a new coordinate system. He described the polar coordinates in his book *Method of Fluxions*.[1] Newton was a prodigious geometer, for in this volume he suggested a total of eight types of coordinate systems. Jacob Bernoulli (1654–1705) was the first to extensively use polar coordinates, publishing his methods in 1691 in the mathematical journal, *Acta Eruditorum*.[2] Jacob Hermann (1678–1733) carried on the ideas of Bernoulli to develop what is essentially our modern system of polar coordinates.

In the Cartesian coordinate system, we define the location of each point with two numbers: the first is the point's distance along the x-axis; while the second is the point's distance along the y-axis. However, in the polar coordinate system we define each point by the length of the line drawn from an origin to the point, and the angle this line makes with the

x-axis, which we now call the polar axis (Figure 96). For example, the point p in Figure 96 is defined by the length r, and the angle θ.

Using a coordinate system defined in terms of an angle θ and a radius r, we can achieve some very beautiful curves. It's possible to translate equations defined in Cartesian coordinates into equivalent equations (achieving the same curve) in polar coordinates. For example, in Cartesian coordinates a circle whose center is at the origin and whose radius is 1 is $x^2 + y^2 = 1$. However, when we change this equation into polar coordinates it becomes simply $r = 1$. This short statement says that for all angles θ, the radius always has a length of just 1 (Figure 97). A 45 degree straight line passing through the origin can be written as just $\theta = 45°$ (Figure 98), although it is customary to express the angles used in polar coordinates as radians and not degrees.

While some Cartesian equations become simpler when translated into polar coordinates, others can become much more complex. The conic curves (parabola, ellipse, and hyperbola) when translated into polar coordinates involve trigonometric functions. For example the parabola $y = x^2/4$ becomes

$$r = \frac{1}{1 - \sin\theta}$$

However, polar coordinates can generate some truly wondrous new curves. For example a limaçon with an inner loop is generated by the polar equation

$$r = 1 + 2 \cdot \cos(\theta)$$

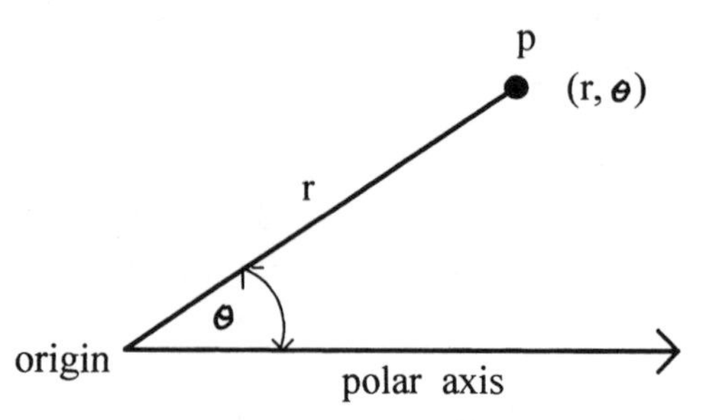

FIGURE 96. Defining the polar coordinate system. Each point is defined by its distance from the polar origin (r) and the angle between the line connecting the point to the origin and the polar axis.

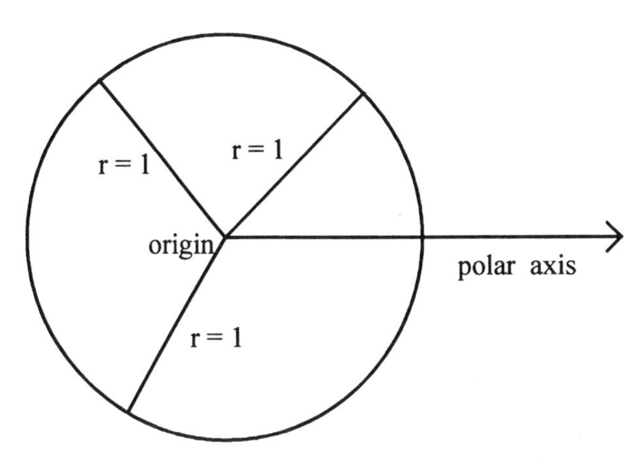

FIGURE 97. The equation for a circle with radius 1 in polar coordinates is simply $r = 1$.

which has been graphed in Figure 99. A rose with four petals is generated by $r = 2 \cdot \cos(2\theta)$ (Figure 100). A lemniscate is the equation $r^2 = 4 \cdot \sin(2\theta)$ (Figure 101).

One of the most beautiful curves ever discovered is the logarithmic spiral which has the polar equation $r = e^{k\theta}$. If we let $k = 0.2$ and θ range from -2π to $+2\pi$, we get the curve in Figure 102. Here again we see the

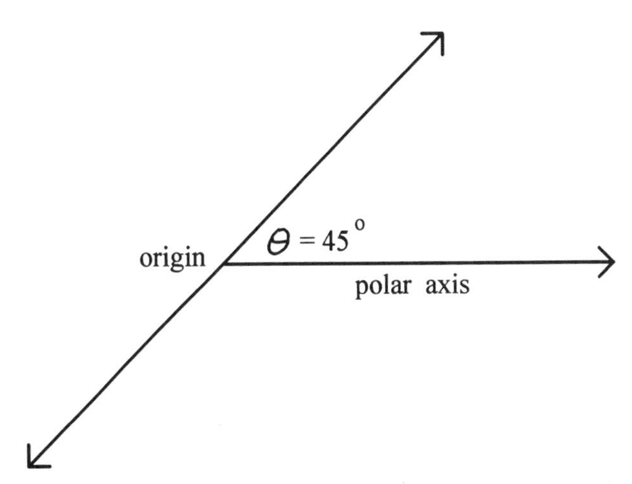

FIGURE 98. The equation for a straight line passing through the origin in polar coordinates at 45 degrees is $\theta = 45°$.

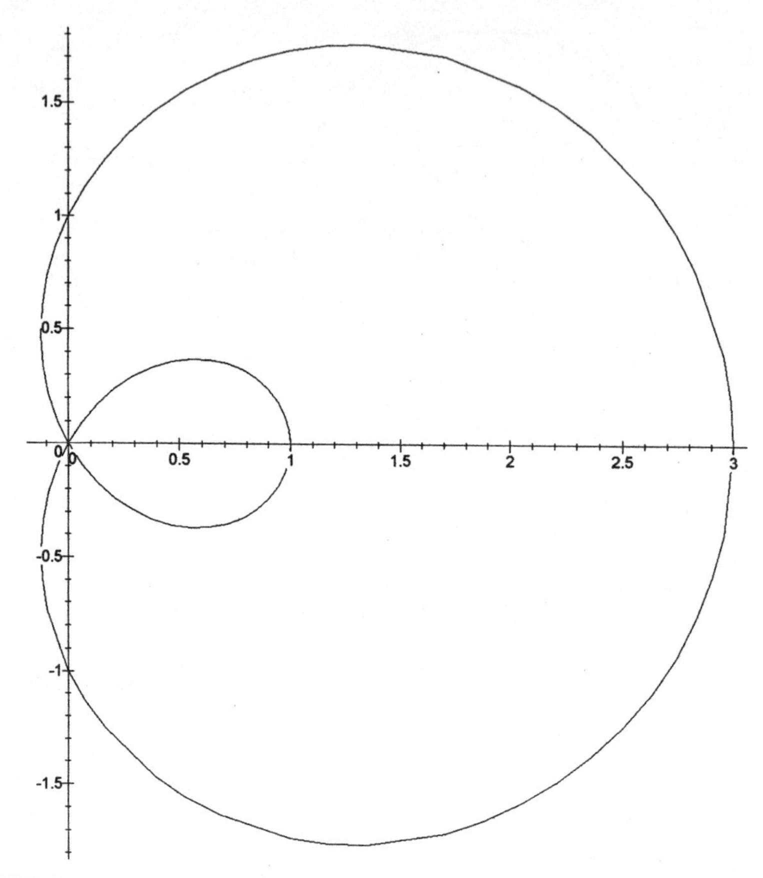

FIGURE 99. In polar coordinates, a limaçon with an inner loop, $r = 1 + 2 \cdot \cos(\theta)$.

mysterious number e showing itself. The logarithmic spiral has some astounding characteristics. For example, it is the only spiral such that a straight line passing through the origin intersects the spiral at each place on the spiral at the same angle (Figure 103). As the curve spirals inward toward the center, it makes an infinite number of complete rotations, each one smaller than the next. Yet, surprisingly, the total length of the spiral from any fixed point inward is finite! This was proved in 1645 by Evangelista Torricelli (1608–1647), who was secretary to Galileo during the last months of Galileo's life.[3] Look at Figure 104 which shows the logarithmic spiral. At point P we find the tangent line to the spiral and extend this

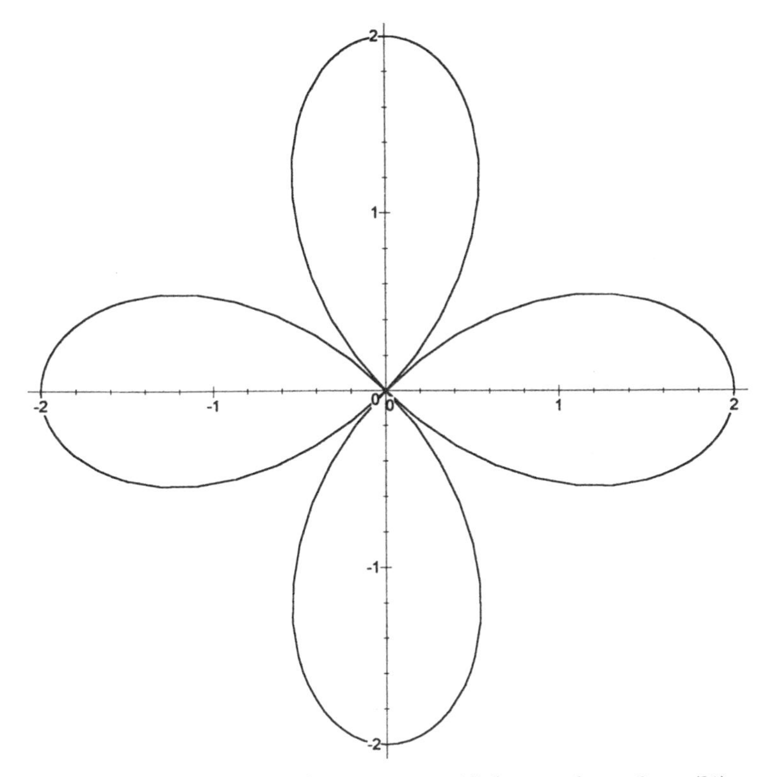

FIGURE 100. In polar coordinates, a rose with four petals, $r = 2 \cdot \cos(2\theta)$.

tangent line until it intersects the y-axis. The length of the line from P to the intersection of the y-axis is exactly equal to the length of the spiral from P to the origin, even though the spiral makes an infinite number of rotations to get there. This is a difficult idea to visualize. Yet, we can get our minds around it if we think of an infinite series of terms adding to a finite amount. Remember the infinite series adding to 1:

$$\sum_{n=1}^{\infty} \frac{1}{2^n} = \frac{1}{2} + \frac{1}{4} + \frac{1}{8} + \frac{1}{16} + \ldots = 1$$

Now think of the length of each spiral turning in toward the origin as one term in another infinite series that adds to a fixed amount, but in this case it adds not to one, but to the length of the tangent line as proved by Torricelli.

We end our glimpse of polar coordinates with an especially beautiful

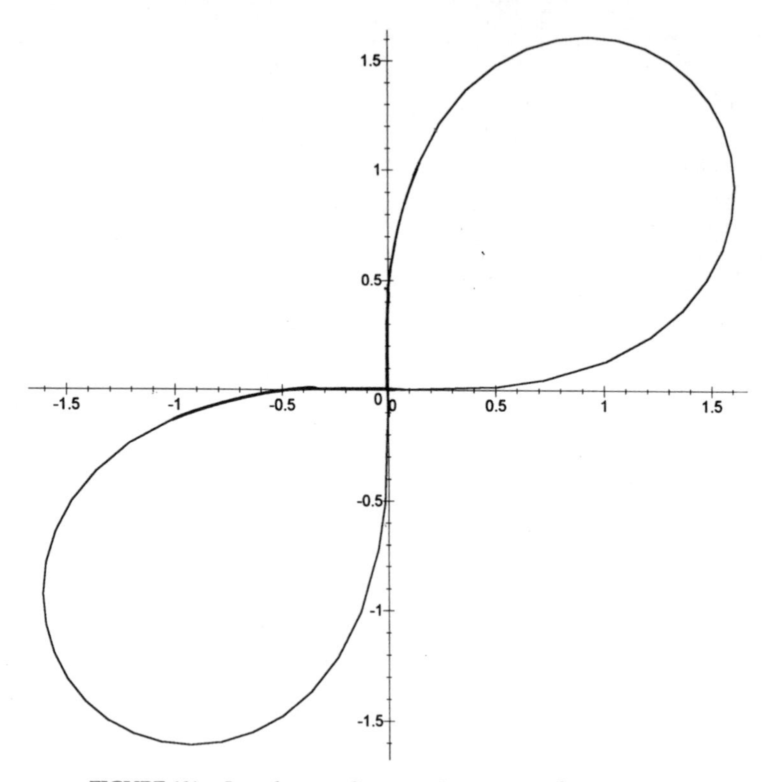

FIGURE 101. In polar coordinates, a lemniscate, $r^2 = 4 \cdot \sin(2\theta)$.

curve, called Henri's Butterfly, generated by the equation, $r = (\sin 4\theta)^2 + \cos 3\theta$, and drawn in Figure 105.

THE MATRIX

We now have the opportunity to introduce an entirely new mathematical object by considering the problem of moving curves from one location on a coordinate system to a new location. Such a move is called a transformation, and transformations lead us to the theory of matrices. The idea behind matrices goes back as far as the Babylonians, 300 B.C., and the Chinese, 200 B.C., who considered arrays of numbers when solving systems of linear equations. Others since then toyed with number arrays when dealing with specific problems but generally did not realize the

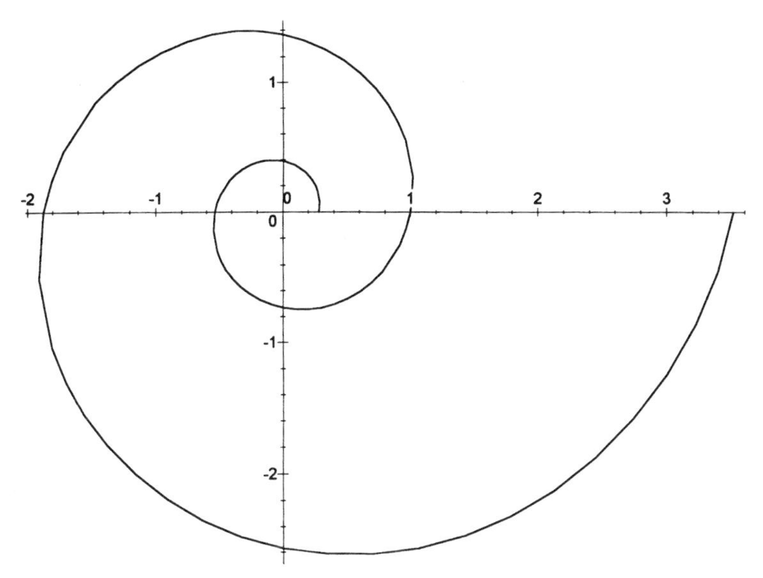

FIGURE 102. In polar coordinates, the beautiful logarithmic spiral, $r = e^{0.2\theta}$.

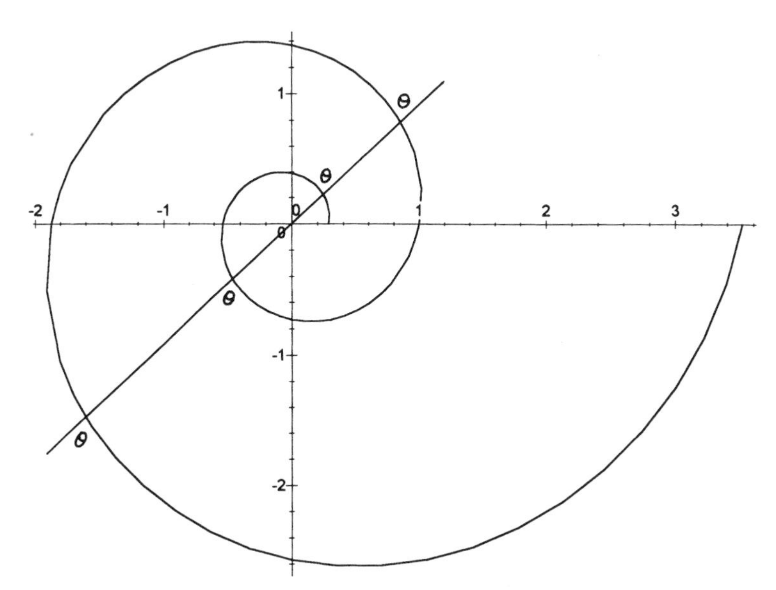

FIGURE 103. A straight line passing through the origin intersects each branch of the logarithmic spiral at the same angle.

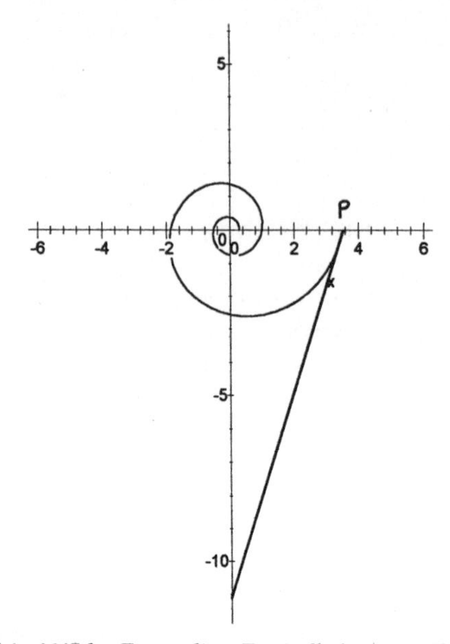

FIGURE 104. Proved in 1645 by Evangelista Torricelli (1608–1647), the length of the logarithmic spiral from point P inward toward the center is exactly equal to the length of the tangent line at point P to its intersection with the y-axis, even though the spiral makes an infinite number of rotations as it circles inward.

grand idea behind their putterings. The two men responsible for developing the modern idea of a matrix are Arthur Cayley (1821–1895) and James Sylvester (1814–1897), two Englishmen who developed not only a close working relationship in mathematics but a deep and lasting personal friendship.

Cayley (Figure 106) studied at Trinity College, Cambridge University, where Isaac Newton had studied and taught some two hundred years before, and where most of England's great mathematicians received their education.[4] Originally, Cayley was slated to become a merchant, but an astute mathematics teacher spotted Arthur's true talents and directed him away from business. He graduated in 1842 and won a three-year fellowship, which allowed him to wander through Europe and think about math while taking nature walks and painting. During this time he produced over two dozen papers on mathematics, but at the conclusion of his fellowship he could find no meaningful teaching position. Thus Arthur Cayley quit mathematics to become a—lawyer!

FIGURE 105. In polar coordinates, Henri's Butterfly, $r = (\sin 4\theta)^2 + \cos 3\theta$.

FIGURE 106. Arthur Cayley (1821–1895). Photograph from Brown Brothers, Sterling, PA.

James Joseph Sylvester was seven years senior to Arthur Cayley and also studied at Cambridge University, but at St. John's College rather than Trinity College. He was a student a half decade before Cayley, and did not connect with the younger man while at Cambridge. Unfortunately, before he could graduate he was required to take a religious oath to the Church of England. Being Jewish, he refused, and was denied his diploma.

He taught physics in London for several years before moving to America, where he experienced a rather traumatic event. Within a few months of being appointed to a position at the University of Virginia, he struck an insulting student with a stick, causing the student to faint. The youth completely recovered, but Sylvester, thinking he'd killed the boy, escaped the school and then fled all the way back to England. There, his mathematical prospects did not improve and he was forced to enter a different field—he became a lawyer!

Fortunately for us, this brought Sylvester into close contact with

Cayley, who was practicing law in the same London courts. They struck up a friendship, which reinvigorated their earlier interests in mathematics. It is curious to think that had Sylvester not run away from the classroom at the University of Virginia, he would have discovered the student was perfectly fine and probably stayed on for the balance of his career. Thus he never would have met Cayley, nor helped Cayley develop matrix theory.

Cayley went on to become a professor of mathematics at Cambridge, while Sylvester, at his somewhat more advanced age, became a professor at Johns Hopkins University, where he founded the first American mathematical journal, *The American Journal of Mathematics*.

But just what marvelous invention (discovery?) did these two men accomplish? To move an object from one location to another on a coordinate system, we can express the new coordinates for both x and y in terms of the old coordinates. Let the new coordinates be x' and y' while the old coordinates are x and y. We show the transformation as:

$$x' = Ax + By$$
$$y' = Cx + Dy$$

The coefficients A, B, C, and D are real numbers. What happens when we change the x and y into x' and y'? We move all points in the xy-plane to a new location in the $x'y'$-plane. For example, suppose that we have the following transformation:

$$x' = x + y$$
$$y' = -2x + y$$

In Figure 107 we have drawn a circle in the xy-plane with its center at the origin and a radius of one. By taking the xy-plane through our transformation as given above, we have moved all the points of the circle to the larger ellipse. Hence, the effect of our transformation has been to stretch the circle into an ellipse and rotate its axis.

After doing the first transformation, suppose we wanted to do a second according to the following transformation equations:

$$x'' = ax' + by'$$
$$y'' = cx' + dy'$$

The question now becomes: how do I do both transformations at once instead of having to do one after the other? To find the correct equations

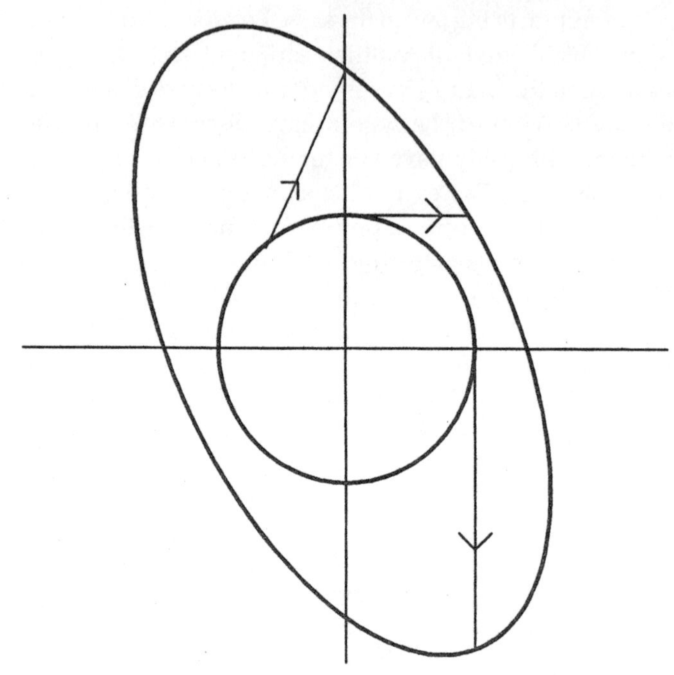

FIGURE 107. A graph of the linear transformation caused by the equations, $x' = x + y$ and $y' = -2x + y$. All the points on the circle have been transformed to the larger ellipse.

which combine the two transformations, we simply substitute the equations for x' and y' into the new set of equations.

$$x'' = a(Ax + By) + b(Cx + Dy)$$
$$y'' = c(Ax + By) + d(Cx + Dy)$$

By multiplying and collecting all the terms we get:

$$x'' = (Aa + bC)x + (aB + bD)y$$
$$y'' = (Ac + Cd)x + (Bc + dD)y$$

But the equations are getting so messy that we are going to get a headache if we don't do some simplification. In our transformations we know the x term of the x' transformation is always in the upper left-hand corner, the y term of the x' equation is in the upper right-hand corner, and so on. Hence we can dispense with all the x and y terms, and write the coefficients in number arrays, or:

$$\begin{pmatrix} a & b \\ c & d \end{pmatrix} \quad \begin{pmatrix} A & B \\ C & D \end{pmatrix} \quad \begin{pmatrix} (aA + bC) & (aB + bD) \\ (Ac + Cd) & (Bc + dD) \end{pmatrix}$$

These three number arrays are what we call matrices. Hence, matrices are simply rectangular arrays of numbers. Each number in the array is called an element of the matrix. Arthur Cayley gave a definition for arithmetic operations with matrices, just as if they were objects similar to numbers. He defined matrix multiplication in such a manner that when we multiply together two matrices that represent transformations, the product turns out to be the transformation that achieves the same result as doing the two transformations in sequence. Therefore, the large matrix above is the product of the two smaller matrices.

Both mathematicians and scientists have discovered that Cayley's definition of matrix multiplication can be applied to the solution of a multitude of applied problems in mathematics and the sciences, and is now a cornerstone of applied mathematics.

Just how is matrix multiplication achieved? First we write the two transformation matrices side by side with the second transformation matrix on the left and the first on the right.

$$\begin{pmatrix} a & b \\ c & d \end{pmatrix} \cdot \begin{pmatrix} A & B \\ C & D \end{pmatrix} = \begin{pmatrix} (aA + bC) & (aB + bD) \\ (Ac + Cd) & (Bc + dD) \end{pmatrix}$$

To see how we actually carry out this multiplication, notice that the upper element on the left-hand side of the product matrix is the combination $(aA + bC)$, which is just the corresponding elements of the top row $(a\ b)$ in the left matrix multiplied by the corresponding elements of the left column of the right matrix added together.

In other words, we took the first element in the first row of the left matrix (a) and multiplied it by the first element of the first column of the right matrix (A) to get the product aA. Then we multiplied the second element of the first row of the left matrix (b) with the second element of the first column of the right matrix (C) to get the product bC. Then we added the two products together: $aA + bC$. This now becomes the first element in the first row and first column of the product matrix.

We compute the other elements in the product matrix in a similar fashion. The second element of row one of the product matrix is $(aB + bD)$, which is the sum of the corresponding elements of the first row of the left matrix multiplied by the elements of the second column of the right matrix. Using this "inner product" rule we compute all the elements. All this multiplying and adding elements sounds extremely confusing, but once we master the technique, the process of carrying out the multiplication is actually quite simple. We gave an example of one transformation as:

$$x' = x + y$$
$$y' = -2x + y$$

Suppose we wanted to do a second transformation which was:

$$x'' = -2x + 3y$$
$$y'' = x - 2y$$

We can compute the total effect of the two transformations by just multiplying the coefficient matrices together, remembering to place the coefficient matrix for the second transformation on the left, or:

$$\begin{pmatrix} -2 & 3 \\ 1 & -2 \end{pmatrix} \cdot \begin{pmatrix} 1 & 1 \\ -2 & 1 \end{pmatrix} = \begin{pmatrix} -8 & 1 \\ 5 & -1 \end{pmatrix}$$

The reader should carry out the multiplication of the two matrices on the left to verify that they yield the product matrix on the right. Now we can directly write the new product transformation equations explicitly as:

$$x'' = -8x + y$$
$$y'' = 5x - y$$

This set of equations will move everything in the plane to the same location as performing the two individual transformations in sequence.

At this point, it would appear that the work of Cayley and Sylvester leads only to a convenient way to write down numbers to do transformations. Yet, their work introduced a new mathematical object (the matrix), which can be manipulated with mathematical operations such as addition, subtraction, and multiplication. We can also define multiplication of a matrix by a real number in the following way:

$$k \cdot \begin{pmatrix} a & b \\ c & d \end{pmatrix} = \begin{pmatrix} ka & kb \\ kc & kd \end{pmatrix}$$

Hence, to multiply any matrix by the real number k, we just multiply every element of the matrix by k. We can define the division of a matrix by a real number as multiplying that matrix by the reciprocal of the real number.

Matrices are like numbers, but they are both more complex than numbers and more powerful. We can think of matrices as the extension of the idea of number just as fractions represent an extension of the integers and complex numbers represent an extension of the real numbers. Each matrix can contain more information than a single number, for it is made up of a set of numbers in specific locations.

Every real number can be written in matrix form. For example, the number 7 can be written as the square matrix:

$$\begin{pmatrix} 7 & 0 \\ 0 & 7 \end{pmatrix}$$

Manipulating this matrix with other matrices is equivalent to manipulating the number 7. Notice in this matrix that two diagonal elements are 0, while the other diagonal elements are both 7. Every real number written as a matrix will have this same form, i.e., a diagonal where all the elements are the number in question with all other elements equal to zero. The number 11 in a matrix of three rows and three columns would be:

$$\begin{pmatrix} 11 & 0 & 0 \\ 0 & 11 & 0 \\ 0 & 0 & 11 \end{pmatrix}$$

We can go even further. Every complex number $a + bi$ can be represented by a matrix of the form:

$$\begin{pmatrix} a & b \\ -b & a \end{pmatrix}$$

To demonstrate how the multiplication of matrices preserves the characteristic of the imaginary unit, i, such that $i^2 = -1$, we can square the matrix representation of i.

$$i \cdot i \rightarrow \begin{pmatrix} 0 & 1 \\ -1 & 0 \end{pmatrix} \cdot \begin{pmatrix} 0 & 1 \\ -1 & 0 \end{pmatrix} = \begin{pmatrix} -1 & 0 \\ 0 & -1 \end{pmatrix} \rightarrow -1$$

On the right we have the matrix representation for the real number -1.

Again, every real and complex number can be written as a matrix, yet by deviating from those restricted forms, the matrix becomes a more universal object. Scientists and mathematicians quickly discovered that by utilizing matrices, they could solve a host of sticky problems, including: rotating axes in analytical geometry; finding solutions to sets of simultaneous linear equations; solving difficult differential equations; attacking problems in electric circuit theory, quantum mechanics, wave propagation, and molecular structure.

SOLVING SIMULTANEOUS EQUATIONS

A great number of applied problems in science can be reduced to a set or system of linear equations. Therefore, understanding how to solve such systems is crucial to unraveling such problems. Let's consider the simple example of solving two simultaneous linear equations. These are the very equations you learned to solve by either substitution or elimination in college algebra. They usually came about when you tried to solve one of those notorious word problems. Using substitution or elimination worked fine when we had just two equations with two unknowns, and it was even possible to use these two techniques when we had three equations and three unknowns. However our teachers generally avoided examples of four or more equations with four or more unknowns, because the two methods we learned back then are not sufficiently strong to solve such large systems. The large number of mathematical operations we must perform when dealing with four or more equations almost guarantees that we will make a mistake, and that mistake will contaminate the final results.

However, using matrix theory we can easily solve large systems of linear equations. We accomplish this by representing the coefficients and constants of the linear system we wish to solve as a matrix. By altering the matrix we change those equations in such a way that the resulting system is easy to solve, yet the solutions to the new system are identical to those of the original system. Each equation in a system of two linear equations is the equation of a line. If we graph each of the linear equations, then the intersection of those two lines represents the solution of the system. Suppose we have the system:

$$3x - 3y = -3$$
$$3x + y = 9$$

Upon inspection we see that the graphs of these two lines cross at approximately the point (2,3), which should be the solution point. While graphing is easy to apply to two equations, this method becomes far too cumbersome as we increase the number of equations. We need a method that yields precise solutions and is relatively easy to apply.

To illustrate this new method, we're going to change the second equation ($3x + y = 9$) so that it becomes a horizontal line, yet intersects the first line in the same place as the original equation. We do this by first subtracting the first equation from the second to get $4y = 12$, or $y = 3$. This is an equation with only one unknown, so it is easy to solve. And the intersection of line $y = 3$ with the first line ($3x - 3y = -3$) is the same point as with the original two equations.

As long as we use a multiple of one equation to subtract or add to another equation, we change the equation without changing where the original two equations intersect. This procedure is called a row operation on a matrix. In addition we can multiply any equation of a system with a constant, and we can interchange any two rows, without changing the solutions for the overall system. In other words, the solution is that point in space where the two lines associated with the equations intersect. Performing row operations allows us to rotate those lines to simplify them while keeping the same point as their intersection.

To see how slick this process of matrix manipulation works for solving systems of linear equations, let's consider a system of three equations and three unknowns.

$$\begin{aligned} x - 2y + z &= 3 \\ 2x + y - 2z &= 1 \\ -x + 3y + z &= -4 \end{aligned}$$

To solve this system using the method of elimination, we would have to take two pairs of equations from the above set of three and eliminate the same unknown from each pair. From this resulting single pair of equations we would eliminate one unknown to solve for the other. Once we had one unknown solved, we would substitute it back into our other equations to solve for the other two unknowns. The whole process would be time consuming, and could easily result in a mistake somewhere along the chain of calculations.

What happens when we move up to larger systems? Say a set of six equations and six unknowns, or ten equations and ten unknowns? Very quickly the methods of elimination and substitution become too unwieldy to use with any confidence that our answers are going to be right.

To solve the system using row operations, we write the equations as a matrix without the x, y, or z, or the equal signs, and we enclose the coefficients and constant terms between large parentheses. This is called the system matrix.

$$\begin{pmatrix} 1 & -2 & 1 & 3 \\ 2 & 1 & -2 & 1 \\ -1 & 3 & 1 & -4 \end{pmatrix}$$

In our system matrix we have used only the coefficients of the three unknowns (x, y, and z) plus the constant terms, the constant terms being placed in a column to the far right.

Now we can begin to have some fun. What we're going to do is to change the above matrix until each successive row has one less unknown represented. This will result in the last row having only the unknown z plus some constant. The value of z can be quickly calculated from the constant. The row directly above the last row will have the unknowns for both y and z. We will have solved for z, so we can substitute z back into this equation and solve for y. We continue this back solving for unknowns until all are solved.

We will know we are done manipulating the system matrix when all the elements of the matrix below the main diagonal from upper left to lower right are zeros. This means we must turn three coefficients in the above matrix into zeros, the 2 in row two and column one, the -1 in row three and column one, and the 3 in row three and column two. The first coefficient to change to zero is the -1 in the lower left corner. The order we use to render the coefficients zero is important, for we must do it in such a manner that we don't take a step backward, i.e., when making one coefficient zero, we don't want another that is already zero to change into something else.

We can make the -1 in the lower left corner a zero by simply adding the first row to the third row. When we do this row operation, we do nothing to the first two rows, even though we are using the first row to adjust the third row. Only the third row changes. Now we carry out the operation. Adding the first row to the third row gives us the four new elements 0, 1, 2, -1. These four new elements are now written in the third row.

$$\begin{pmatrix} 1 & -2 & 1 & 3 \\ 2 & 1 & -2 & 1 \\ -1 & 3 & 1 & -4 \end{pmatrix} \rightarrow \begin{pmatrix} 1 & -2 & 1 & 3 \\ 2 & 1 & -2 & 1 \\ 0 & 1 & 2 & -1 \end{pmatrix}$$

Notice in the altered matrix on the right that the first two rows are unchanged because we were only bringing about a change in the third row. The third row now has a zero coefficient for its x term.

We now wish to make the x coefficient in row two a zero. This is achieved by multiplying each element in the first row by -2 and adding the result to the corresponding elements in the second row. Multiplying the top row by -2 yields $-2, 4, -2$, and -6. Adding these numbers to the corresponding elements in the second row gives us $0, 5, -4, -5$, which now becomes row two.

$$\begin{pmatrix} 1 & -2 & 1 & 3 \\ 2 & 1 & -2 & 1 \\ 0 & 1 & 2 & -1 \end{pmatrix} \rightarrow \begin{pmatrix} 1 & -2 & 1 & 3 \\ 0 & 5 & -4 & -5 \\ 0 & 1 & 2 & -1 \end{pmatrix}$$

We only have one more element to change to zero, the 1 in the second column, third row. However, the only row we can use to change this to a zero is the second row which contains a 5 in second column. We could multiply the second row by a $-1/5$ and add to the third row, but this would involve some very messy fractions which we wish to avoid. Therefore we will now swap the positions of the second and third rows, which is one of our legitimate row operations. This places a 1 in the second column of the second row and a 5 in the second column of the third row.

$$\begin{pmatrix} 1 & -2 & 1 & 3 \\ 0 & 5 & -4 & -5 \\ 0 & 1 & 2 & -1 \end{pmatrix} \rightarrow \begin{pmatrix} 1 & -2 & 1 & 3 \\ 0 & 1 & 2 & -1 \\ 0 & 5 & -4 & -5 \end{pmatrix}$$

Now we will multiply each element in the second row by a -5 and add the result to the corresponding elements in the third row.

$$\begin{pmatrix} 1 & -2 & 1 & 3 \\ 0 & 1 & 2 & -1 \\ 0 & 5 & -4 & -5 \end{pmatrix} \rightarrow \begin{pmatrix} 1 & -2 & 1 & 3 \\ 0 & 1 & 2 & -1 \\ 0 & 0 & -14 & 0 \end{pmatrix}$$

We now have the system matrix in the form we want, for all three elements below the main diagonal are 0. Notice that the last row contains only a -14 and 0. We can now write this as the equation $-14z = 0$. Solving for z we get $z = 0$.

The second row of our new systems matrix can be interpreted as $y + 2z = -1$. But we know that z is 0, so we substitute 0 for z into the equation and discover that $y = -1$. With our two unknowns in hand, we solve the first row for x, or $x = 1$.

While the row operations on a matrix may seem daunting when first encountered, they are quickly mastered, and actually turn out to be fun to perform. My students soon discover the seduction matrix operations have over me when they ask for help on their homework. I start to show them the row operations, fully intending to stop and allow them to finish the process on their own. But something strange comes over me, and before I can stop, I've manipulated the entire matrix and solved the problem.

IDENTITY AND INVERSE MATRICES

Matrices are objects which are similar to numbers but do not share all the characteristics of numbers. For example, real and complex numbers obey the commutative law under multiplication. That is, if A and B are numbers, then $A \cdot B = B \cdot A$. This is a law we learn so early in our education that we simply take it for granted that 2×3 is the same as 3×2. However, matrices do not have this characteristic. If A and B are square matrices then usually $A \cdot B \neq B \cdot A$. In arithmetic, an analogous situation is when we consider the operation of subtraction between two numbers. For example, $7 - 2 \neq 2 - 7$.

If we consider a matrix to be any rectangular array of numbers, then multiplication is not defined between every pair of matrices. We can only multiply two matrices A and B when the number of rows of A equals the number of columns of B. Otherwise multiplication is not possible. It is significant that matrices do not share the commutative law of multiplication. The algebra of the complex numbers is an example of the most general algebra possible obeying all the fundamental laws of arithmetic, including the commutative law.[5]

In our discussion of matrices we have shown that they can be added, subtracted, and (in certain situations) multiplied. Yet, we have not mentioned division. To understand matrix division we must first define the inverse of a matrix. A square matrix is one whose number of columns equals the number of rows. Therefore, any two square matrices of the same size can be multiplied. The square matrix where all the elements are zeros except for the main diagonal, which are ones, is the identity matrix, and we designate it as I. This is because whenever we multiply an identity matrix with another square matrix of the same size we get the original matrix back again. Hence we have:

$$A \cdot I = I \cdot A = A$$

Thus, for square matrices the identity matrix, I, works just the way the identity element, 1, works for the real numbers. If n is a real number, then $1 \cdot n = n \cdot 1 = n$.

Using the idea of an identity matrix we can now define an inverse matrix. B is the inverse matrix to A if the following holds: $A \cdot B = I$. That is, by multiplying a matrix by its inverse, we get the identity matrix. The inverse of A is generally written as A^{-1}. This is analogous to real numbers where we multiply a number by its reciprocal or $A(1/A) = 1$, e.g., $3(1/3) = 1$. However there is a significant difference between matrices and numbers. Not every square matrix has an inverse. Matrices which have no inverse we called singular. If a matrix does have an inverse we call it a nonsingular matrix.

Now we can define matrix division. If matrix C is divided by matrix A, then we are looking for a matrix, B, which when multiplied by A yields C, or $A \cdot B = C$. In other words, $C/A = B$ corresponds to $A \cdot B = C$. But what is B? To find out what B is, we multiply each side of the equation $A \cdot B = C$ by the inverse of A, or

$$A^{-1} \cdot A \cdot B = A^{-1} \cdot C$$

But the inverse of A multiplied by A is just the identity matrix, or $I \cdot B = A^{-1} \cdot C$. However, the identity matrix times any matrix is the original matrix, or $I \cdot B = B = A^{-1} \cdot C$. This means that the matrix B we are after is just the inverse of A multiplied by C. Hence, division is defined in terms of multiplication, and for A to divide C, then A must have an inverse. This is analogous to real numbers where we can define the division of 5 by 7 as 5 multiplied by the inverse of 7 or $5/7 = 5 \cdot (1/7)$.

We have now entered the realm of matrix algebra where we must be very cautious, since we know that matrices do not always act in the same way as their parents, the numbers. However, using the idea of an inverse and the identity matrix we can develop another faster method to solve large systems of linear equations. In modern science and technology, it is not uncommon to encounter problems we know how to solve, i.e., we know the general techniques that will work, but because of the sheer size of the problems, the number of unknowns and equations, we simply can't apply our techniques to get meaningful answers. This is especially true of simultaneous linear equations which are inordinately large. Therefore, the techniques involving matrix algebra greatly increase our power over these problems by allowing us to handle much larger systems.

Suppose we write only the coefficients of our original system of linear equations without the constant column. This will give us a square matrix. Next we write the unknowns as a column matrix with the elements x, y, and z. Then we can write the column of constants as its own column matrix which will be three rows but only one column. We can now equate these three matrices in the following manner:

$$\begin{pmatrix} 1 & -2 & 1 \\ 2 & 1 & -2 \\ -1 & 3 & 1 \end{pmatrix} \cdot \begin{pmatrix} x \\ y \\ z \end{pmatrix} = \begin{pmatrix} 3 \\ 1 \\ -4 \end{pmatrix}$$

Notice that if we multiply the two matrices on the left of the equal sign we get the expression for the unknowns of our original system, while the right side is the expression for the constants of our system. We are now going to do a little matrix algebra so that the solutions for x, y, and z pop out. First, let the three matrices above be represented as A for the coefficient matrix, U for the unknowns matrix, and C for the constant matrix. Thus we have $A \cdot U = C$. Next we find the inverse of A, which we designate as A^{-1}. We can now multiply each side of the equation with our inverse matrix to get:

$$A^{-1} \cdot A \cdot U = A^{-1} \cdot C$$

But the expression $A^{-1} \cdot A$ is just the identity matrix or $A^{-1} \cdot A = I$. Thus we have:

$$A^{-1} \cdot A \cdot U = I \cdot U = A^{-1} \cdot C.$$

However, the identity matrix times any other matrix is equal to the other matrix, or

$$U = A^{-1} \cdot C$$

Now the left side of the equation is nothing more than the unknowns matrix. This means that if we can find an inverse for matrix A and multiply that inverse times the constant matrix, the answers to our system of linear equations will jump right out. This reduces the whole process to just finding the inverse to a square matrix.

Fortunately for us, any system of linear equations with a unique solution for the unknowns will have a corresponding coefficient matrix which has an inverse. We can think of a system of equations with n unknowns as an n-dimensional space, and each equation is a subspace. The solution is that point where all the subspaces intersect. If such a point exists, then the original matrix has an inverse. All we have to do is to figure out some way to find it. We can do this by using row operations. Suppose we have the following system of three equations and three unknowns:

$$\begin{aligned} x + \ y - 2z &= -3 \\ -x + \ y + \ z &= \ 4 \\ 2x - 2y + \ z &= \ 1 \end{aligned}$$

First we create the following set of three matrices:

$$\begin{pmatrix} 1 & 1 & -2 \\ -1 & 1 & 1 \\ 2 & -2 & 1 \end{pmatrix} \cdot \begin{pmatrix} x \\ y \\ z \end{pmatrix} = \begin{pmatrix} -3 \\ 4 \\ 1 \end{pmatrix}$$

Now our chore is to find the inverse to the three-by-three matrix on the left and multiply it by the column matrix on the right. To compute the inverse of the matrix we create a new matrix, which is the conjunction of the matrix for which we want to find the inverse and the identity matrix. In our problem we join the three-by-three matrix on the left above with a three-by-three identity as follows:

$$\begin{pmatrix} 1 & 1 & -2 & 1 & 0 & 0 \\ -1 & 1 & 1 & 0 & 1 & 0 \\ 2 & -2 & 1 & 0 & 0 & 1 \end{pmatrix}$$

We have created a three-by-six matrix with the original matrix in the left three columns and the identity matrix in the right three columns. We now use row operations to transform the left three-by-three matrix above into an identity matrix. This will magically transform the identity matrix on the right into our inverse matrix. To carry out this transformation, we must take the six elements that are off the diagonal of the left matrix and make them all zero. We then divide each row so that the elements on the diagonal are all ones. We will do this by first changing the three elements below the diagonal to zeros. The first to be changed is the 2 in the third row, first column. To achieve this we multiply the first row by -2 and add the result to the third row to get

$$\begin{pmatrix} 1 & 1 & -2 & 1 & 0 & 0 \\ -1 & 1 & 1 & 0 & 1 & 0 \\ 2 & -2 & 1 & 0 & 0 & 1 \end{pmatrix} \to \begin{pmatrix} 1 & 1 & -2 & 1 & 0 & 0 \\ -1 & 1 & 1 & 0 & 1 & 0 \\ 0 & -4 & 5 & -2 & 0 & 1 \end{pmatrix}$$

Notice now that the three-by-three identity matrix on the right is no longer an identity matrix for it has the element -2 in the third row.

We now change the -1 in the second row, first column into a zero by simply adding the first row to the second row to yield:

$$\begin{pmatrix} 1 & 1 & -2 & 1 & 0 & 0 \\ -1 & 1 & 1 & 0 & 1 & 0 \\ 0 & -4 & 5 & -2 & 0 & 1 \end{pmatrix} \rightarrow \begin{pmatrix} 1 & 1 & -2 & 1 & 0 & 0 \\ 0 & 2 & -1 & 1 & 1 & 0 \\ 0 & -4 & 5 & -2 & 0 & 1 \end{pmatrix}$$

The next element to change to zero is -4 in the third row. We accomplish this by multiplying the second row by 2 and adding to the third row:

$$\begin{pmatrix} 1 & 1 & -2 & 1 & 0 & 0 \\ 0 & 2 & -1 & 1 & 1 & 0 \\ 0 & -4 & 5 & -2 & 0 & 1 \end{pmatrix} \rightarrow \begin{pmatrix} 1 & 1 & -2 & 1 & 0 & 0 \\ 0 & 2 & -1 & 1 & 1 & 0 \\ 0 & 0 & 3 & 0 & 2 & 1 \end{pmatrix}$$

We have succeeded in changing the three elements below the diagonal to zeros and we must do the same for the three elements above the diagonal. The first to be changed will be the -2 in the first row, third column. We do this by multiplying the second row by -2 and adding the results to the first row:

$$\begin{pmatrix} 1 & 1 & -2 & 1 & 0 & 0 \\ 0 & 2 & -1 & 1 & 1 & 0 \\ 0 & 0 & 3 & 0 & 2 & 1 \end{pmatrix} \rightarrow \begin{pmatrix} 1 & -3 & 0 & -1 & -2 & 0 \\ 0 & 2 & -1 & 1 & 1 & 0 \\ 0 & 0 & 3 & 0 & 2 & 1 \end{pmatrix}$$

We're almost there; only two more elements to go. Next we change the -1 in the second row, third column to a zero by multiplying the third row by ⅓ and adding to the second row. This introduces fractions into our calculations, but they will present only a small increase in difficulty.

$$\begin{pmatrix} 1 & -3 & 0 & -1 & -2 & 0 \\ 0 & 2 & -1 & 1 & 1 & 0 \\ 0 & 0 & 3 & 0 & 2 & 1 \end{pmatrix} \rightarrow \begin{pmatrix} 1 & -3 & 0 & -1 & -2 & 0 \\ 0 & 2 & 0 & 1 & \frac{5}{3} & \frac{1}{3} \\ 0 & 0 & 3 & 0 & 2 & 1 \end{pmatrix}$$

We next change the -3 in the first row, second column into a zero by multiply the second row by 3/2 and adding to the first row.

$$\begin{pmatrix} 1 & -3 & 0 & -1 & -2 & 0 \\ 0 & 2 & 0 & 1 & \frac{5}{3} & \frac{1}{3} \\ 0 & 0 & 3 & 0 & 2 & 1 \end{pmatrix} \rightarrow \begin{pmatrix} 1 & 0 & 0 & \frac{1}{2} & \frac{1}{2} & \frac{1}{2} \\ 0 & 2 & 0 & 1 & \frac{5}{3} & \frac{1}{3} \\ 0 & 0 & 3 & 0 & 2 & 1 \end{pmatrix}$$

We're just about home. All we have to do is divide the second row by 2 and the third row by 3, which will change the three-by-three matrix on the left into a true identity matrix.

$$\begin{pmatrix} 1 & 0 & 0 & \frac{1}{2} & \frac{1}{2} & \frac{1}{2} \\ 0 & 2 & 0 & 1 & \frac{5}{3} & \frac{1}{3} \\ 0 & 0 & 3 & 0 & 2 & 1 \end{pmatrix} \rightarrow \begin{pmatrix} 1 & 0 & 0 & \frac{1}{2} & \frac{1}{2} & \frac{1}{2} \\ 0 & 1 & 0 & \frac{1}{2} & \frac{5}{6} & \frac{1}{6} \\ 0 & 0 & 1 & 0 & \frac{2}{3} & \frac{1}{3} \end{pmatrix}$$

The right hand three-by-three matrix has now been transformed into the inverse of our original matrix. With this matrix we can find our solution by multiplying it by the column matrix of our constants.

$$\begin{pmatrix} \frac{1}{2} & \frac{1}{2} & \frac{1}{2} \\ \frac{1}{2} & \frac{5}{6} & \frac{1}{6} \\ 0 & \frac{2}{3} & \frac{1}{3} \end{pmatrix} \cdot \begin{pmatrix} -3 \\ 4 \\ 1 \end{pmatrix} = \begin{pmatrix} 1 \\ 2 \\ 3 \end{pmatrix}$$

Therefore, our solutions are $x = 1$, $y = 2$, and $z = 3$. The reader can also verify that the inverse matrix satisfies the relationship $A \cdot A^{-1} = I$ by multiplying the original matrix by its inverse, or

$$\begin{pmatrix} 1 & 1 & -2 \\ -1 & 1 & 1 \\ 2 & -2 & 1 \end{pmatrix} \cdot \begin{pmatrix} \frac{1}{2} & \frac{1}{2} & \frac{1}{2} \\ \frac{1}{2} & \frac{5}{6} & \frac{1}{6} \\ 0 & \frac{2}{3} & \frac{1}{3} \end{pmatrix} = \begin{pmatrix} 1 & 0 & 0 \\ 0 & 1 & 0 \\ 0 & 0 & 1 \end{pmatrix}$$

Of course, we have gone through a lot of effort to solve three equations in three unknowns. Why would anyone want to learn matrix algebra and the tedious inverses if we can solve such a system with substitution or elimination? When we use matrix algebra to solve large systems of equations, we use a computer, for computers love to perform row operations and to find inverses to matrices. Therefore, to solve large systems of equations we program computers to take the inverse, and the system is conquered.

Many years ago while a student at the University of Utah, I worked for the Bureau of Economic and Business Research. We were doing a two-year study of Utah's economy by building an input–output table. To construct such a table required that we divide Utah's economy into thirty-nine sectors. Then we interviewed 1200 businesses belonging to those sectors to find out how much each sector of the state's economy bought and sold to each of the other sectors. When we were done we had a large table with all thirty-nine sectors listed across the top and also listed down the left side. Each intersection of the table showed the total value of goods and services sold by the row sector to the column sector.

From such an input–output table we needed to learn how much the overall economy of the state would change by making a change in one or more sectors. To compute this change it was necessary to treat the table as a matrix and find its inverse. To manually take the inverse of such a large matrix would have been impossible. Instead, we used the university's computer, and the inverse was found in a matter of hours. Today, with faster computers, we could find the answer in minutes.

MATRIX STRANGENESS

Of course finding the solutions to systems of linear equations is not the only use for matrix algebra. By combining the idea of a matrix with other functions we can generate some truly exotic mathematical creatures. A standard problem in geometry is the rotation of axes. For example, if we have a function whose graph is tilted in an unsatisfactory manner, we can rotate the axis through the angle ϕ with the following transformation matrix:

$$\begin{pmatrix} \cos\phi & \sin\phi \\ -\sin\phi & \cos\phi \end{pmatrix}$$

Here we have combined trigonometric functions with a matrix. To carry out the rotation, we simply solve the following equation.

$$\begin{pmatrix} x' \\ y' \end{pmatrix} = \begin{pmatrix} \cos\phi & \sin\phi \\ -\sin\phi & \cos\phi \end{pmatrix} \cdot \begin{pmatrix} x \\ y \end{pmatrix}$$

This is analogous to solving the following system of linear equations:

$$\begin{aligned} x' &= x \cdot \cos\theta + y \cdot \sin\theta \\ y' &= -x \cdot \sin\theta + y \cdot \sin\theta \end{aligned}$$

We can also combine square matrices with both the trigonometric functions and infinite series in the following two beautiful expressions, where A is a square matrix:

$$\sin(A) = A - \frac{A^3}{3!} + \frac{A^5}{5!} - \ldots + (-1)^r \frac{A^{2r+1}}{(2r+1)!} + \ldots$$

$$\cos(A) = I - \frac{A^2}{2!} + \frac{A^4}{3!} - \ldots + (-1)^r \frac{A^{2r}}{(2r)!} + \ldots$$

Each matrix term in the infinite series is well defined because we know that a square matrix can be multiplied by itself to yield another square matrix of the same size. Hence, we can raise any square matrix to any positive integer power. We also know that we can divide any matrix by a real number.

Using our identity for e and the sine and cosine functions we also have the following identity for a square matrix A:

$$e^{iA} = \cos(A) + i \cdot \sin(A)$$

With both the infinite series and the exponential identity the final object turns out to be a square matrix.

Thus we realize that the amazing things we could do with real and complex numbers, such as using trigonometric functions and infinite series, can also be done with matrices. Yet, we have only scratched the surface of this wondrous field of mathematics, for there is much more that can be accomplished using the sophisticated form of the matrix.

RIEMANN'S GEOMETRY

So far we have considered only two-dimensional curves in a two-dimensional coordinate system. However, it would be impossible to do satisfactory physics and celestial mechanics without working in the third dimension. One of the most important characteristics of two-dimensional Euclidean space is that the Pythagorean theorem holds everywhere. In fact we use the Pythagorean theorem to calculate the distance between any two points in the Cartesian plane. Figure 108 shows two points, one located at the origin and the second with coordinates (x, y). We can form the right triangle with the distance between the two points as the triangle's hypotenuse, s. Now the question becomes: What are the lengths of the two legs? From the coordinates of the two points we notice that the horizontal leg of the triangle is simply x, and that the length of the vertical leg is y. From the Pythagorean theorem we have $s^2 = x^2 + y^2$. To find the distance, s, between the two points we simply take the square root of the right-hand side. This expression defines the metric on our space and guarantees that it is flat or Euclidean. A flat space has no or zero curvature.

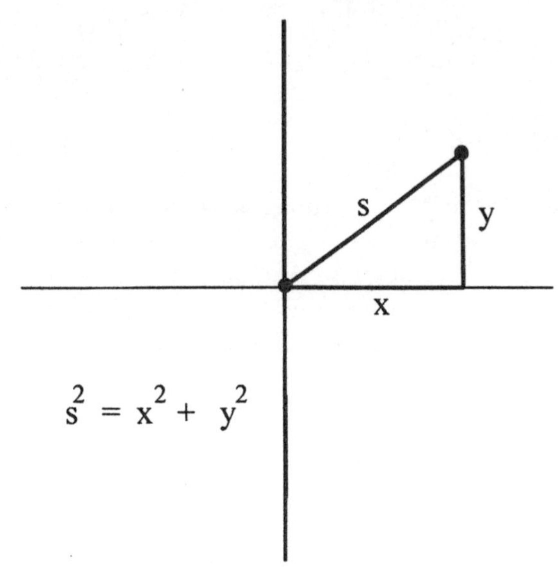

FIGURE 108. The metric on two-dimensional "flat" or Euclidean space is defined by the Pythagorean theorem, $s^2 = x^2 + y^2$. A flat space has no or zero curvature.

When we go to three dimensions, we must consider the expression for the distance between any two points. Figure 109 shows a three-dimensional coordinate system with three axes: x-axis, y-axis, and z-axis. Every point within our space is defined by its three coordinates (x, y, z). We have placed our first point at the origin and the second point at the coordinates (x, y, z). What is the expression for the length of the line, s, that connects them?

If we drop a line straight down from our point (x, y, z) we hit the point $(x, y, 0)$. Since this point is now in the two-dimensional plane of xy, we can compute its distance s_{xy} (the dotted line) from the origin to our point $(x, y, 0)$ with the Pythagorean theorem, or $s_{xy} = \sqrt{x^2 + y^2}$. However, this line is one leg of the right triangle defined by the points $(0, 0, 0)$, (x, y, z), and $(x, y, 0)$. The other leg is simply the length z. Hence, using the Pythagorean theorem again we have:

$$s^2 = (\sqrt{x^2 + y^2})^2 + z^2, \text{ or } s^2 = x^2 + y^2 + z^2$$

This shows that when going from two dimensions to three dimensions, the metric which defines the curvature of Euclidean space is changed only by adding the additional z^2 term. But what about spaces that

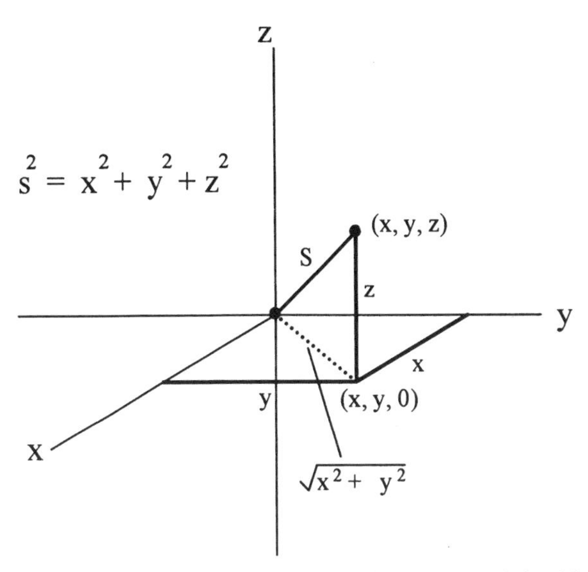

FIGURE 109. The metric on three-dimensional Euclidean space is defined by the Pythagorean theorem, $s^2 = x^2 + y^2 + z^2$.

have a curvature? For example, the surface of a sphere (e.g., Earth's surface) is not flat, and the normal laws of Euclidean geometry don't apply. This is easy to see when we consider the fact that in Euclidean geometry the interior angles of a triangle always add to 180 degrees. However, that is not true on Earth's surface. Form a triangle using two longitudes from the north pole to the equator. The two longitudes plus the equator form a triangle whose interior angles add to more than 180 degrees since the longitudes intersect the equator at right angles.

To answer this question we must turn to the brilliant mathematician, Georg Friedrich Bernhard Riemann (1826–1866). Riemann was one of six children raised by his widowed father, a Lutheran pastor with a meager income. Riemann was supposed to follow his father into the ministry, but his fear of public speaking made him dread commanding the pulpit. He begged his father for permission to change his major to that of mathematics. Receiving approval, Riemann studied at both the University of Berlin and then went to the University of Göttingen where he became a student of the famous Carl Gauss.

In 1854 Riemann became a professor at Göttingen, at which time he was required to deliver an introductory lecture for the faculty. His paper entitled, "On the Hypotheses Which Lie at the Foundation of Geometry," may well have been the most influential lecture ever given in mathematics.[6] The notion that other geometries existed besides Euclidean geometry had already been determined, for the Russian mathematician Nicolai Ivanovitch Lobachevsky (1793–1856) had delivered a paper as early as 1826 on the characteristics of at least one non-Euclidean geometry.[7] Yet, non-Euclidean geometries were still considered somewhat in the backwaters of mathematics.

Riemann's paper changed all that. He proposed that geometry not be thought of as collections of points and lines, but as sets of ordered n-tuplets and the rules for determining the distances between elements in the set. An n-tuplet is simply an ordered set of n numbers. For example, in two-dimensional Euclidean space, every point is uniquely defined by an ordered pair of real numbers (x, y). In three-dimensional space every point is uniquely defined by three real numbers (x, y, z). In n-dimensional space, each point has a unique address given by an n-tuplet $(x, y, z, \ldots, n)$.

The rules for determining the distances between elements in the set are the geometry's metric, and they define space's curvature. Classical geometry studied curves and surfaces in their entirety, but modern geometry was concerned with the microscopic shape of space that surrounded each point of the space. The infinitesimal distance between points in normal three-dimensional Euclidean space was given by the formula $ds^2 = dx^2 + dy^2 + dz^2$, where the small d reminds us we are working at the infinitesimal level. Riemann generalized the idea of geometry as the study of possible manifolds, rather than any rigidly defined metric. In the broadest sense, a manifold is nothing more than a collection of objects of a set. Geometry then becomes the study of the conditions placed upon the objects of sets, rather than implying the characteristics of space by visualizing the shape of the space. Riemann was not interested in just the study of three-dimensional space or even four-dimensional space, but the more general characterization of n-dimensional spaces. Thus his approach was much more universal than much of the work before.

Riemann's expression for the most general distance formula in three-dimensional space is the nine terms in the following equation[8]:

$$ds^2 = g_{11}dx^2 + g_{12}dxdy + g_{13}dxdz + g_{21}dydx + g_{22}dy^2 + g_{23}dydz + g_{13}dzdx + g_{23}dzdy + g_{33}dz^2$$

where the various coefficients, g_{ij}, are either constants or functions of x, y, and z. If we let $g_{11} = g_{22} = g_{33} = 1$ while all the other coefficients are 0, then we have the distance formula for three-dimensional Euclidean space with no curvature. If we move up one dimension to four-dimensional space, then Riemann's distance formula contains not nine coefficients, but sixteen.

However, we can see at once that Riemann's long expression, called the Riemann metric, holds the potential for defining an infinite number of different spaces with different curvatures. In fact, Einstein used Riemann's idea to develop his general theory of relativity. Einstein was the first to realize the importance of the speed of light as a constant in transformation equations describing real space–time, since the speed of light is always constant, regardless of the reference system. Just how was the speed of light to be combined with the transformation equations? Hermann Minkowski (1864–1909), one of Einstein's teachers, gave the mathematical answer. He began with three axes of space: x-axis, y-axis, and z-axis.[9] Then he attached to these a fourth axis of time, which is designated as ict, where i is the imaginary unit ($\sqrt{-1}$), c is the speed of light, and t is time. This becomes the Minkowski space. Speed is distance per unit time. If we multiply speed by time we end up with simply distance, which agrees with the meaning of the other three axes, yet incorporates time into this fourth axis.

I can't draw you a picture of how this fourth axis is attached to the other three, because I simply can't visualize in four dimensions. In flat Euclidean space, the Minkowski space has a distance formula which is:

$$s^2 = x^2 + y^2 + z^2 - c^2t^2$$

Notice that in this expression the i term is missing because when we square i we get -1. This explains the negative in front of the c^2t^2 term. Now Einstein introduced curvature into this Minkowski space, a curvature caused by the presence of matter—gravitation. Hence, matter is creating the curvature of space and this is what we interpret as gravity. The distance formula now becomes a huge expression with sixteen different coefficients.[10] Thus Einstein used not only Gauss' idea for curved coordinate systems, and Riemann's expression for the distance formula, but also the imaginary axis as his time axis. Thus we have a mathematical description of how matter curves space.

Nevertheless, Riemann not only has the distinction of delivering possibly the most profound lecture on geometry, but he repeated his accomplishment by delivering in 1858 equally profound eight-page paper on number theory. Whether he could have produced a third such thunderous paper the world will never know, for he died tragically from tuberculosis when he was only 39.

8

Extending the Form

In the previous chapters we saw the rich earth we can till by extending addition into the infinite and creating infinite series. We can perform the same magic with other elementary math operations, including products and fractions. Infinite series involve the addition and/or subtraction of an infinite number of terms, and then investigating whether the resulting series converge (tend toward a finite value) or diverge. We can extend this process of the infinite for both multiplication and division. First we'll consider multiplication.

INFINITE PRODUCTS

As with the series, we have a special symbol to represent a product, i.e., the result of multiplying numbers together. With the series we used capital sigma (Σ). To indicate a product we use capital pi (Π), which must not be confused with the familiar symbol, small pi (π), used for the ratio of the circumference of a circle to its diameter. Capital pi is entirely distinct and represents a set of terms all multiplied together.

Suppose we wish to multiply the first six odd numbers together. Using capital pi we can show it as:

$$\prod_{n=1}^{6} (2n - 1) = 1 \cdot 3 \cdot 5 \cdot 7 \cdot 9 \cdot 11 = 10{,}395$$

In the above equation n is our index and n ranges from 1 to 6. When $n = 1$, $(2n - 1)$ yields 1, and with each successive increase in n we get $(2n - 1)$ to yield the next odd number. Capital pi indicates we are going to multiply these six odd numbers together to get 10,395.

However, to achieve a really remarkable breakthrough we must extend this idea of successive multiplications to infinitely many factors. Leonard Euler, whom we have already met in relation to infinite series, was the first to give detailed attention to infinite products. As with infinite series, infinite products can either diverge or converge to some finite amount. Consider the simple infinite product consisting of all ones.

$$\prod_{n=1}^{\infty} 1_n = 1 \cdot 1 \cdot 1 \cdot 1 \cdot \ldots = 1$$

This is just a fancy way of saying that if we take an infinite number of ones and multiply them all together, we still end up with the quantity one. A simple infinite product that diverges is:

$$\prod_{n=1}^{\infty} n = 1 \cdot 2 \cdot 3 \cdot 4 \cdot \ldots = \infty$$

because the product gets arbitrarily large as we multiply with more and more factors. The question now becomes: Which infinite products converge, and what do they converge to?

Many very lovely infinite products converge to interesting and exotic values. Several, as you might suspect, involve the value of π. For example, the following beautiful expression was derived by John Wallis (1616–1703), who is considered the most outstanding English mathematician before Isaac Newton. Wallis wrote two influential books on mathematics, used mathematical principles to decode secret Royalist messages for the Parliamentarians during the English Civil War, and later became one of the founders of the English Royal Society. For fifty years he held a chair in geometry at Oxford College, Cambridge.

$$\prod_{n=1}^{\infty} \frac{(2n)^2}{(2n-1)(2n+1)} = \frac{2}{1} \cdot \frac{2}{3} \cdot \frac{4}{3} \cdot \frac{4}{5} \cdot \frac{6}{5} \cdot \frac{6}{7} \cdot \frac{8}{7} \cdot \frac{8}{9} \cdot \ldots = \frac{\pi}{2}$$

There are two striking things about Wallis' infinite product: first, the pattern is so simple and elegant. Second, the product is half of π. Can anything be more astounding—that an infinite multiplication can generate a value equal to half the ratio of the circumference of a circle to its diameter?

An earlier infinite product involving π was derived by Francois Viète (1540–1603), a Frenchman who was trained as a lawyer and worked as a French politician, but fortunately found enough time to do serious mathematics. He published several math books and, like John Wallis, mastered cryptography, deciphering secret coded messages for the French Crown during the war with Spain in 1590.

$$\frac{\sqrt{2}}{2} \cdot \frac{\sqrt{2+\sqrt{2}}}{2} \cdot \frac{\sqrt{2+\sqrt{2+\sqrt{2}}}}{2} \cdot \frac{\sqrt{2+\sqrt{2+\sqrt{2+\sqrt{2}}}}}{2} \cdot \ldots = \frac{2}{\pi}$$

This is truly an exceptional expression for it is really an infinite series imbedded within an infinite product, and it contains only the number 2 combined with the idea of a square root.

A third infinite product involving π similar to the first expression is:

$$\prod_{n=1}^{\infty} \frac{(4n)^2}{(4n-1)(4n+1)} = \frac{4}{3} \cdot \frac{4}{5} \cdot \frac{8}{7} \cdot \frac{8}{9} \cdot \frac{12}{11} \cdot \frac{12}{13} \cdot \frac{16}{15} \cdot \frac{16}{17} \cdot \ldots = \frac{\pi \cdot \sqrt{2}}{4}$$

Leonhard Euler discovered one of the most astonishing infinite products that connects all natural numbers (the positive integers) to all prime numbers. As we've mentioned, prime numbers are especially elusive to mathematicians because their distribution within the natural numbers appears so random. Euclid proved over two thousand years ago that an infinite number of primes exist, yet we know of no elementary formula that identifies each successive prime nor can we quickly determine if a large number is prime or not. Today, many mathematicians consider one of the most profound unsolved problems within mathematics to be the Riemann Hypothesis, a conjecture by Georg Riemann about the distribution of primes. Therefore, Euler's discovered infinite product, called the Zeta Function, is especially delicious for it is based on the infinity of primes and their connection to the number sequence.

$$\sum_{n=1}^{\infty} \frac{1}{n^s} = \prod_{p=\text{primes}}^{\infty} \frac{1}{1 - \frac{1}{p^s}}$$

Notice on the left we have the infinite series of the reciprocals of all natural numbers raised to a specific power, s. On the right we have a strange product comprising all the primes raised to the same power, s. For Euler, s could be any real number, but Riemann's extension of the Zeta Function defines it for all complex numbers.

A second way of writing the right-hand side of the above expression gives more of an intuitive feel for its pattern.

$$\sum_{n=1}^{\infty} \frac{1}{n^s} = \prod_{p=\text{prime}}^{\infty} \frac{p^s}{p^s - 1}$$

We can now see that the right side is simply the product of each prime raised to a power s, divided by the same number minus one. Suppose we let $s = 2$. This makes the expression on the left the infinite sum of the reciprocals of all square numbers which we already know is $\pi^2/6$. Hence we have a most profound equation:

$$\sum_{n=1}^{\infty} \frac{1}{n^2} = \prod_{p=\text{prime}}^{\infty} \frac{p^2}{p^2 - 1} = \frac{2^2}{(2^2 - 1)} \cdot \frac{3^2}{(3^2 - 1)} \cdot \frac{5^2}{(5^2 - 1)} \cdot \frac{7^2}{(7^2 - 1)} \cdot \ldots = \frac{\pi^2}{6}$$

When I see this equation it gives me goosebumps. It has everything: an infinite series, an infinite product, all natural numbers, all prime numbers, and finally the splendid constant π. This equation does everything! It butters your toast and fixes your car. Let's compute the first few terms of the product to see how quickly it converges to $\pi^2/6$. Using the first twenty-five primes below one hundred we get:

$$\prod_{p=\text{prime}}^{97} \frac{p^2}{(p^2 - 1)} = \frac{2^2}{(2^2 - 1)} \cdot \frac{3^2}{(3^2 - 1)} \cdot \ldots \cdot \frac{97^2}{(97^2 - 1)} = 1.641945\ldots$$

Calculating $\pi^2/6$ to six decimal places right of the decimal point yields 1.644934. Hence, our product of only the first twenty-five primes was accurate to within 0.2 percent of the value of $\pi^2/6$—remarkable!

We can actually express both sides as their reciprocals or:

$$\prod_{p=\text{prime}}^{\infty} \left(1 - \frac{1}{p^2}\right) = \left(1 - \frac{1}{2^2}\right) \cdot \left(1 - \frac{1}{3^2}\right) \cdot \left(1 - \frac{1}{5^2}\right) \cdot \left(1 - \frac{1}{7^2}\right) \cdot \ldots = \frac{6}{\pi^2}$$

If we let $s = 1$, then the left side of Euler's equation becomes the harmonic series, which we know diverges. The above expressions only converge for real values of s greater than 1.

It is important to know when infinite products converge and diverge. For purposes of consistency, mathematicians say that an infinite product also diverges if its limit is zero or $\prod f(n) = 0$. Therefore, for an infinite product to converge, it must converge to some finite value different than zero. This means that the individual factors of an infinite product, F_n, tend to 1 as n increases or:

$$\text{If } \prod_{n=1}^{\infty} F_n \text{ converges then } \lim_{n\to\infty} F_n = 1$$

Since the individual factors tend to 1, we can rewrite any converging infinite product as:

$$\prod_{n=1}^{\infty} (1 + a_n)$$

where the terms a_n must approach zero. Now, the terms a_n can be either all positive or negative. This gives us a nice theorem regarding converging infinite products:

> *Theorem*: The infinite products $\Pi (1 + a_n)$ and $\Pi (1 - a_n)$ converge if and only if the infinite series Σa_n converges.

The above theorem establishes a nice relationship between converging series and converging products. For example, we can ask whether the infinite product $\Pi (1 + 1/n)$ converges or diverges. We know that the infinite series $\Sigma (1/n)$ is the harmonic series and that it diverges. This means that the corresponding product, $\Pi (1 + 1/n)$ must diverge also, as well as the product $\Pi (1 - 1/n)$, which will diverge to zero. A second theorem regarding the convergence of infinite products is:

> *Theorem*: The infinite product ΠF_n converges if and only if the infinite sum $\Sigma \ln (F_n)$ converges, where ln represents the natural logarithm.

This is a nice theorem and easy to appreciate when we write out the terms of the infinite series involving the logarithm.

$$\sum_{n=1}^{\infty} \ln (F_n) = \ln (F_1) + \ln (F_2) + \ln (F_3) + \ldots$$

Now we use the general logarithm rule that says:

$$\ln (A) + \ln (B) = \ln (A \cdot B)$$

We can combine the infinity of logarithmic terms into just the logarithm of an infinite product or:

$$\sum_{n=1}^{\infty} \ln (F_n) = \ln (F_1) + \ln (F_2) + \ldots = \ln (F_1 \cdot F_2 \cdot F_3 \cdot \ldots) = \ln \left(\prod_{n=1}^{\infty} F_n \right)$$

Hence, the infinite series of logarithms of F_n equals the logarithm of the infinite products of F_n.

We just learned that every infinite series $\Sigma (1/n^s)$ converges when s is greater than 1. Hence, every infinite product of the form $\Pi (1 + 1/n^s)$ and $\Pi (1 - 1/n^s)$ will converge to a value other than zero when $s > 1$. This tells us that the infinite products of the following forms all converge: $\Pi (1 + 1/n^2)$, $\Pi (1 - 1/n^2)$, $\Pi (1 + 1/n^3)$, and $\Pi (1 - 1/n^3)$, etc.

We can even have an infinite product where the signs alternate between successive factors. For example the infinite series $1 - 1/3 + 1/5 - 1/7 + \ldots$ converges to $\pi/4$ and therefore the associated product must also converge.

$$\prod_{n=1}^{\infty}\left(1+\frac{(-1)^{n+1}}{2n-1}\right)=\left(1+\frac{1}{1}\right)\cdot\left(1-\frac{1}{3}\right)\cdot\left(1+\frac{1}{5}\right)\cdot\left(1-\frac{1}{7}\right)\cdot\left(1+\frac{1}{9}\right)\cdot\ldots=\sqrt{2}$$

Again we have an amazing expression. The numerator of the second term in the product, $(-1)^{n+1}$ is used just to reverse the sign in each successive term. The denominator, $2n-1$, produces successive odd numbers. Of course the surprising thing about the product is that it is exactly equal to $\sqrt{2}$.

Once we have looked at the alternating reciprocals of the odd numbers, as above, it is only natural to ask what happens when we consider the infinite product $\Pi(1 - 1/n^2)$, because we know that the infinite series $\Sigma(1/n^2)$ converges to $\pi^2/6$. Therefore, the associated product must converge to something. It is an easy exercise to show that this product converges to ½ when we let n begin with an index value of 2 rather than a value of 1, for if we begin with 1 the first term would be 0.

$$\prod_{n=2}^{\infty}\left(1-\frac{1}{n^2}\right)=\left(1-\frac{1}{2^2}\right)\cdot\left(1-\frac{1}{3^2}\right)\cdot\left(1-\frac{1}{4^2}\right)\cdot\left(1-\frac{1}{5^2}\right)\cdot\ldots=\frac{1}{2}$$

Now we want to ask what the infinite product $\Pi(1 + 1/n^2)$ converges to. However, this requires a little more explaining. We have already pointed out many of the special qualities of the number e, including the graph of the function e^x which describes many decay and growth rates found in nature. If we graph e^x we get the solid curve on Figure 110, and if we graph the function e^{-x} we get the dashed curve in Figure 110, which is just the e^x curve reflected around the y-axis. If we add the values of e^x and e^{-x} together and divide by 2 (i.e., take their arithmetic mean) then we get the curve shown in Figure 111. This curve and its associated function, $(e^x + e^{-x})/2$, is called the hyperbolic cosine function. By changing the sign to a negative, we get the hyperbolic sine function, $(e^x - e^{-x})/2$. These two curves and their associated functional expressions have proven to be very useful in mathematics and the sciences. The curve made when a heavy cord or cable is suspended from two points of equal height looks much like a parabola. At one time it was believed that the cables from suspension

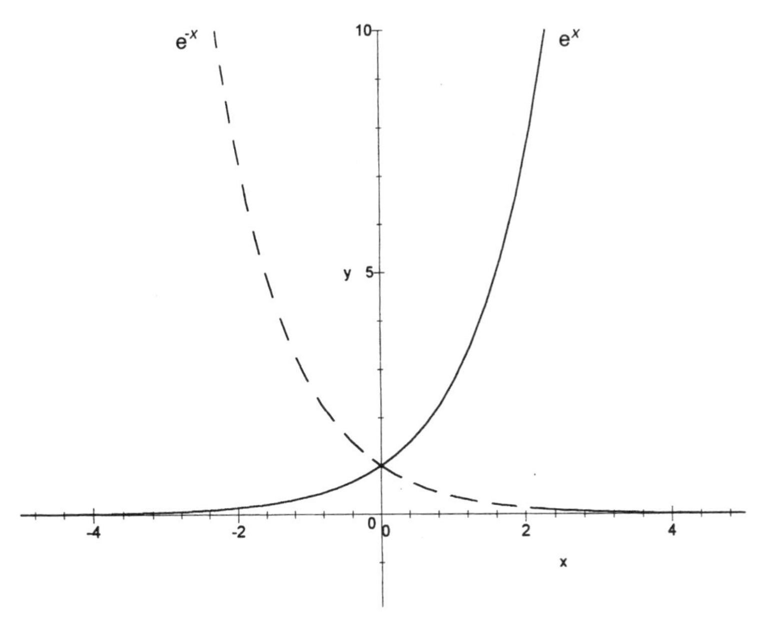

FIGURE 110. The Exponential Function, e^x, and its associated function, e^{-x}.

bridges had the shape of parabolas. We now know that such shapes are catenary curves and have the following form of the hyperbolic cosine:

$$y = a \cdot \cosh(\theta / a)$$

The hyperbolic functions are usually shown as simply $\sinh(x)$ and $\cosh(x)$, where the h distinguishes them from the trigonometric functions of $\sin(x)$ and $\cos(x)$.

$$\sinh(x) = \frac{e^x - e^{-x}}{2} \qquad \cosh(x) = \frac{e^x + e^{-x}}{2}$$

Our infinite product under consideration does not converge to something simple such as ½, but to the remarkable expression:

$$\prod_{n=2}^{\infty}\left(1 + \frac{1}{n^2}\right) = \left(1 + \frac{1}{2^2}\right)\cdot\left(1 + \frac{1}{3^2}\right)\cdot\left(1 + \frac{1}{4^2}\right)\cdot \ldots = \frac{\sinh(\pi)}{\pi} = \frac{e^{\pi} - e^{-\pi}}{2\pi} \approx 3$$

Here we see another example of a truly astounding mathematical form. We have a simple infinite product that converges to a value, which includes

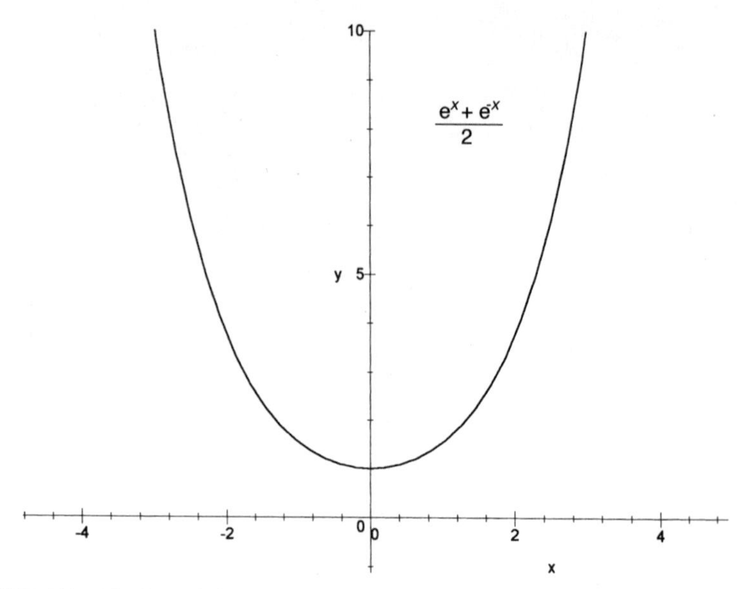

FIGURE 111. The hyperbolic cosine function, $(e^x + e^{-x})/2$ is the arithmetic mean of the Exponential Function, e^x, and its associated function e^{-x}. The hyperbolic sine function is defined as $(e^x - e^{-x})/2$. Such functions describe hanging chains and cables.

not only the number e but also the number π. This same infinite product is associated with an infinite series, namely $\Sigma\,(1/n^2)$. This in turn is a specific instance of the Zeta function, and therefore itself equal to an infinite product including all the prime numbers: $\Pi\,[p^2/(p^2 - 1)]$. So much within mathematics is so incredibly connected.

We have illustrated several infinite products containing π, and we can also add some beautiful equations consisting of the constant e. A previously mentioned infinite series involving the natural logarithm of 2 is:

$$\frac{1}{1} - \frac{1}{2} + \frac{1}{3} - \frac{1}{4} + \frac{1}{5} - \frac{1}{6} + \ldots = \ln(2) \approx 0.693147\ldots$$

We can manipulate the above infinite series using the properties of logarithms and exponents to get the following exotic expression:

$$\left(\frac{e^1}{e^{1/2}}\right) \cdot \left(\frac{e^{1/3}}{e^{1/4}}\right) \cdot \left(\frac{e^{1/5}}{e^{1/6}}\right) \cdot \left(\frac{e^{1/7}}{e^{1/8}}\right) \cdot \left(\frac{e^{1/9}}{e^{1/10}}\right) \cdot \ldots = 2$$

We can also give an infinite product for just e.

$$\frac{2}{1}\cdot\left(\frac{4}{3}\right)^{1/2}\cdot\left(\frac{6\cdot 8}{5\cdot 7}\right)^{1/4}\cdot\left(\frac{10\cdot 12\cdot 14\cdot 16}{9\cdot 11\cdot 13\cdot 15}\right)^{1/8}$$
$$\left(\frac{18\cdot 20\cdot 22\cdot 24\cdot 26\cdot 28\cdot 30\cdot 32}{17\cdot 19\cdot 21\cdot 23\cdot 25\cdot 27\cdot 29\cdot 31}\right)^{1/16}\cdot\ldots = e$$

Notice the elegant symmetry and progression for this expression. Consecutive even numbers appear in the numerators, while consecutive odd numbers appear in the denominators. The number of terms within each set of parentheses always doubles as do the denominators of the exponents.

Since we have shown the infinite products pertaining to hyperbolic sine and hyperbolic cosine, it is only fair that we show infinite products for both the trigonometric sine and cosine functions.

$$\theta\cdot\prod_{n=1}^{\infty}\left(1-\frac{\theta^2}{n^2\pi^2}\right)=\theta\cdot\left(1-\frac{\theta^2}{\pi^2}\right)\cdot\left(1-\frac{\theta^2}{2^2\pi^2}\right)\cdot\left(1-\frac{\theta^2}{3^2\pi^2}\right)\cdot\ldots=\sin(\theta)$$

$$\prod_{n=1}^{\infty}\left(1-\frac{4\theta^2}{(2n-1)^2\pi^2}\right)=\left(1-\frac{4\theta^2}{\pi^2}\right)\cdot\left(1-\frac{4\theta^2}{3^2\pi^2}\right)\cdot\left(1-\frac{4\theta^2}{5^2\pi^2}\right)\cdot\ldots=\cos(\theta)$$

where θ is in radians.

Not all infinite products of the form $\prod(1 \pm c/n^k)$, where c and k are constants, depend on complicated expressions. For example:

$$\prod_{n=3}^{\infty}\left(1-\frac{4}{n^2}\right)=\left(1-\frac{4}{3^2}\right)\cdot\left(1-\frac{4}{4^2}\right)\cdot\left(1-\frac{4}{5^2}\right)\cdot\ldots=\frac{1}{6}$$

Both the hyperbolic and trigonometric functions provide a long list of interesting infinite products, but we will offer only a few more examples. The infinite product $\prod(1 + 1/n^3)$ provides an arresting expression involving both the hyperbolic cosine and π.

$$\prod_{n=1}^{\infty}\left(1+\frac{1}{n^3}\right)=\left(1+\frac{1}{1^3}\right)\cdot\left(1+\frac{1}{2^3}\right)\cdot\left(1+\frac{1}{3^3}\right)=\frac{\cosh\left(\frac{\pi\cdot\sqrt{3}}{2}\right)}{\pi}=$$
$$\frac{e^{(\pi\sqrt{3})/2}+e^{(-\pi\sqrt{3})/2}}{2\pi}\approx 2.428$$

Our next example of an infinite product is stunning in its simplicity and elegance, encompassing not only the infinite product by the limit concept, the number e, and all prime numbers.

$$\lim_{n\to\infty}\left(\prod_{p=\text{prime}}^{p\leq n} p\right)^{1/n} = e$$

This is such an astounding assertion, we must compute the first few terms to see if they really do progress toward e.

when $n = 2$ $2^{1/2} = 1.414\ldots$
when $n = 3$ $(2\cdot 3)^{(1/3)} = 6^{(1/3)} = 1.817\ldots$
when $n = 5$ $(2\cdot 3\cdot 5)^{(1/5)} = 30^{(1/5)} = 1.974\ldots$
when $n = 7$ $(2\cdot 3\cdot 5\cdot 7)^{(1/7)} = 210^{(1/7)} = 2.146\ldots$

It appears that we're approaching the value of e, or 2.71828 . . . . Let's jump ahead and compute the function for all primes less than 100. When $n = 97$, $(\Pi\, p)^{1/n} = 2.370685\ldots$. We're getting closer, but approaching the true value of e rather slowly now.

The following is a strange mixture of an infinite product and an infinite nested radical. For any positive real number x we have:

$$x = \sqrt{x\cdot\sqrt{x\cdot\sqrt{x\cdot\sqrt{x\cdot\ldots}}}}$$

A more general expression of this identity is:

$$\sqrt[n]{x} = \sqrt[n+1]{x\cdot\sqrt[n+1]{x\cdot\sqrt[n+1]{x\cdot\ldots}}}$$

The next example of an infinite product comes to us by way of the MathSoft Web Page, which contains many fine mathematical formulas and identities. Suppose we begin with a circle which has a radius of one unit length, and then inscribe within this circle an equilateral triangle as in Figure 112. We then inscribe another circle within that triangle. Inside the second circle we inscribe a square followed by another circle, followed by a pentagon. We continue this process, each time inscribing a circle followed by a regular polygon with one more side. Surprisingly, the figures do not continue inward toward the center of the first circle, but approach a smaller circle as a limit. This is because each successive polygon has an additional side. The inscribed polygons approach the shape of a circle. What is the radius of the limiting circle?

$$\text{radius} = \prod_{n=3}^{\infty}\cos\left(\frac{\pi}{n}\right) = \cos\left(\frac{\pi}{3}\right)\cdot\cos\left(\frac{\pi}{4}\right)\cdot\cos\left(\frac{\pi}{5}\right)\cdot\ldots\approx 0.11494\ldots$$

We leave the infinite products with a strange expression involving both e and matrices.

FIGURE 112. Beginning with a circle of radius one, polygons with increasing numbers of sides are alternately inscribed within circles. The radius of the circle that is the limit to this process is $\Pi \cos(\pi/n)$, which is approximately 0.11494 . . . .

$$\prod_{n=1}^{\infty} \begin{pmatrix} \frac{1}{n} & 1 \\ 0 & 1 \end{pmatrix} = \begin{pmatrix} \frac{1}{k!} & \Sigma \frac{1}{k!} \\ 0 & 1 \end{pmatrix} \cdot \prod_{n=k+1}^{\infty} \begin{pmatrix} \frac{1}{n} & 1 \\ 0 & 1 \end{pmatrix} = \begin{pmatrix} 0 & e \\ 0 & 1 \end{pmatrix}$$

In the middle term we have shown the multiplication of the first k matrices and how the elements of the product matrix evolve. The upper left element becomes the reciprocal of $k!$ while the upper right term becomes the expression for the sum of these reciprocals. As we take n to infinity, $1/k!$ tends to zero, while $\Sigma\, 1/k!$ tends to e. Remarkable, simply remarkable.

INFINITE CONTINUED FRACTIONS

We are now ready to move on to our next elementary form: the continued fraction. In simplest terms, a continued fraction is a fraction which contains another fraction in either the numerator and/or the denominator. If the additional fractions always appear in the numerators, such as the one below, we say the continued fraction is in ascending order.

$$5 + \cfrac{2 - \cfrac{1 + \cfrac{4}{7}}{3}}{9}$$

However, if all additional fractions appear in the denominators we say the continued fraction is in descending order, as in the following example,

$$2 - \cfrac{1}{4 + \cfrac{2}{5 - \cfrac{3}{9}}}$$

Continued fractions in descending order are of special interest because they are related to the exotic transcendental numbers. In the case where all the numerators are equal to 1 the fraction is called a simple continued fraction. For example:

$$4 + \cfrac{1}{2 + \cfrac{1}{7 - \cfrac{1}{6}}}$$

A shorthand method for writing simple continued fractions is to write the first term, a_0, which is an integer and not part of the fraction proper, followed by a semicolon, and then all the denominators in order, separated by commas, all within brackets, or $[a_0; a_1, a_2, a_3, \ldots, a_n]$. Hence the above continued fraction would be written as [4; 2, 7, 6]. Finite continued fractions such as [4; 2, 7, 6] always represent rational numbers, i.e., numbers that can be written as a proper fraction, p/q where p and q are whole numbers. We are really interested in extending the idea of a simple continued fraction into the infinite. Do such fractions have meaning, and do they converge to definite values? Such a fraction would be:

$$1 + \cfrac{1}{2 + \cfrac{1}{2 + \cfrac{1}{2 + \ldots}}}$$

where the ellipses indicate that the form continues without end. We can write the short version of the above infinite continued fraction as [1; 2, 2, 2,

. . .]. If the infinite continued fraction has denominators that repeat themselves infinitely in a pattern, then we show this by placing a dot above those numbers that are to be repeated, just as we do when we show infinite repeating decimals. Hence, [1; 2, 2, 2, . . .] becomes simply $[1; \dot{2}]$.

Continued fractions may have been first used by the Indian mathematician, Aryabhata the Elder (476–550), who used them to solve linear indeterminate equations.[2] The Italian mathematician Pietro Antonio Cataldi (1548–1626) used them to study the square roots of numbers.[3] Cataldi was educated in Bologna and began teaching mathematics at the tender age of seventeen. During his life he wrote thirty books, working in the fields of number theory, algebra, and geometry. Constructing an infinite continued fraction for a square root is an illustrative exercise and is a way to see how useful they can be in approximating square roots.

First we begin with a simple identity: $\sqrt{2} = \sqrt{2}$. Now we add and subtract 1 from the right side to get: $\sqrt{2} = 1 + (\sqrt{2} - 1)$. Our next step is to write the term within the parentheses as a fraction and then multiply its numerator and denominator by its conjugate.

$$\sqrt{2} = 1 + \frac{(\sqrt{2} - 1)(\sqrt{2} + 1)}{(\sqrt{2} + 1)}$$

Collecting all the terms in the numerator we get:

$$\sqrt{2} = 1 + \frac{1}{(\sqrt{2} + 1)}$$

We now rewrite the denominator to get the original term $(\sqrt{2} - 1)$, or:

$$\sqrt{2} = 1 + \frac{1}{2 + (\sqrt{2} - 1)}$$

We now repeat the whole process by writing the term within the parentheses as a fraction and multiplying both the numerator and denominator by its conjugate. This yields the wonderful repeating infinite continued fraction:

$$\sqrt{2} = 1 + \cfrac{1}{2 + \cfrac{1}{2 + \cfrac{1}{2 + \cfrac{1}{2 + \dots}}}}$$

What would the simplest infinite continued fraction (a fraction where all the numerators and denominators are equal to one) be equal to? We have already given this secret away. It is just the Golden Mean, or $\phi = (1 + \sqrt{5})/2 = [1; \dot{1}]$.

We can, in fact, create a simple infinite continued fraction for every square root (quadratic surd). For example:

$$\sqrt{3} = [1; 1, 2, 1, 2, 1, 2, \ldots] = [1; \dot{1}, \dot{2}]$$
$$\sqrt{5} = [2; 4, 4, 4, \ldots] = [2; \dot{4}]$$
$$\sqrt{7} = [2; 1, 1, 1, 4, 1, 1, 1, 4, \ldots] = [2; \dot{1}, \dot{1}, \dot{1}, \dot{4}]$$

The amazing thing about infinite continued fractions is that every algebraic number, i.e., a number that represents the solution of a polynomial with rational coefficients, has an infinite continued fraction expansion that is unique and is built from an infinite repeating pattern. This is analogous to the decimal expansion of rational numbers where every rational number has either a terminating decimal or an infinite repeating decimal.

But algebraic numbers, which include the integers, fractions, and surds, make up only a puny fraction of the number line. The great bulk of numbers on the number line are the strange transcendental numbers, such as π and e. The algebraic numbers form an infinite set that is countable, while the transcendental numbers form a set that is infinitely larger—uncountable. If we could dump all the real numbers from the number line into a large barrel and then randomly withdraw just one, we are guaranteed what kind of number that would be—it would be transcendental, for the proportion that are algebraic is so minuscule that the chances of drawing one is, for all practical purposes, zero.

What does the simple infinite continued fraction expansion look like for a transcendental number? The continued fraction for e is the whopping:

$$e = 2 + \cfrac{1}{1 + \cfrac{1}{2 + \cfrac{1}{1 + \cfrac{1}{1 + \cfrac{1}{4 + \cfrac{1}{1 + \cfrac{1}{1 + \cfrac{1}{6 + \cfrac{1}{1 + \cfrac{1}{1 + \cfrac{1}{8 + \ldots}}}}}}}}}}}$$

The more compact expression is $e = [2; 1, 2, 1, 1, 4, 1, 1, 6, 1, 1, 8, \ldots]$. No pattern repeats itself infinitely, for every third integer (2, 4, 6, and 8) increases by 2. Yet, it is enough of a pattern to tell us what the next three integers will be. For transcendental numbers, their simple infinite continued fractions are never infinitely repeating. Hence, the continued fraction representation of numbers offers a way to distinguish between algebraic numbers and transcendental numbers. The following theorem guarantees that every real number, both algebraic and transcendental, has a unique continued fraction.

> *Theorem*: Every real number can be written as a unique simple continued fraction. The fraction is finite in length when the real number is rational, infinite and repeating when the real number is irrational and algebraic, and infinite and nonrepeating when the real number is transcendental.

An example of a continued fraction involving π is the lovely:

$$\frac{\pi}{4} = \cfrac{1}{1 + \cfrac{1^2}{2 + \cfrac{3^2}{2 + \cfrac{5^2}{2 + \cfrac{7^2}{2 + \cfrac{9^2}{2 + \ldots}}}}}}$$

In this example, the continued fraction is not simple because the numerators are not all equal to one. Yet, it is not a repeating fraction even though all the denominators after the 1 are 2, because the numerators keep changing. But its pattern is discernable at once, for the numerators are the sequence of squared odd numbers.

We have not considered the question of when simple infinite continued fractions converge, and when they diverge. For this we have the very nice theorem:

> *Theorem*: For the simple infinite continued fraction, $[a_0; a_1, a_2, a_3, \ldots]$, to converge, it is necessary and sufficient that the series $\Sigma\, a_n$ diverge.[4]

This means that if the denominators of a simple continued fraction diverge, then the fraction itself converges; if the denominators converge, then the continued fraction must diverge. The theorem demonstrates that a beautiful relationship exists between infinite continued fractions and infinite series. Again we encounter the mysterious connections within mathematics. It is a truly startling idea, for it says that associated with every diverging infinite series is a number that represents the convergence of its associated continued fraction.

We must also consider how to determine the value of an infinite continued fraction when it has an infinite repeating pattern. For short repeating patterns, this is an elementary problem, but when the pattern becomes longer, it can be daunting. Consider the following simple repeating continued fraction, [3; 3, 3, 3, . . .]. Because the denominators form a diverging series, i.e., $\Sigma\, 3 = \infty$, we know that the continued fraction equals some real number, L, or:

$$L = 3 + \cfrac{1}{3 + \cfrac{1}{3 + \cfrac{1}{3 + \dots}}}$$

Now look at the large denominator under the uppermost division line. This expression is identical to the whole continued fraction. Therefore, if the entire continued fraction is equal to L, then this denominator is also equal to L and we can substitute L to get: $L = 3 + 1/L$. Now we multiply through by L to get the quadratic equation, $L^2 = 3L + 1$. When we solve this quadratic equation we get two answers, $3/2 + \sqrt{13}/2$ and $3/2 - \sqrt{13}/2$. The second answer is negative and makes no sense in this application, because the continued fraction is clearly positive. Therefore, the value of our continued fraction is just $3/2 + \sqrt{13}/2$, or approximately 3.30277 . . . .

We can generalize the above procedure and solve the infinite repeating continued fraction of form $[a; \dot{b}]$ to get:

$$[a; b] = a - \tfrac{1}{2}b + \tfrac{1}{2}\sqrt{(b^2 + 4)}$$

Using the above formula, the infinite continued fraction $[4; \dot{6}]$ has the value of $1 + \sqrt{10}$. To solve for repeating continued fractions with periods exceeding one term is significantly more difficult. For example, if we have a period of two, as in the fraction $[a; \dot{b}, \dot{c}]$, then the value of the fraction becomes:

$$[a; \dot{b}, \dot{c}] = ab - \tfrac{1}{2}bc + \tfrac{1}{2}\sqrt{bc(bc + 4)}$$

Constructing infinite continued fractions for specific values can be both fun and challenging. However, the most valuable knowledge is knowledge about all continued fractions. This next example reveals a startling fact about the great majority of such fractions, and is due to the work of the Soviet mathematician, Aleksandr Yakovlevich Khinchin (1894–1959).[5] Consider the general continued expansion for an arbitrary real number r:

$$r = a_0 + \cfrac{1}{a_1 + \cfrac{1}{a_2 + \cfrac{1}{a_3 + \dots}}}$$

Now let's form the geometric mean of the first n denominators: $a_1, a_2, a_3, \dots a_n$. The geometric mean (GM) of n terms is just the nth root of their product or $\mathrm{GM} = (a_1 \cdot a_2 \cdot a_3 \cdot \dots \cdot a_n)^{1/n}$. We now want to know what GM becomes as we consider more and more terms (n grows very large). Because every real number has a unique simple continued fraction expansion, either finite (rational numbers) or infinite, one would expect that the value of GM varies significantly for different real numbers. However, Khinchin proved that for almost all real numbers the following remarkable identity holds:

$$\lim_{n\to\infty} (a_1 \cdot a_2 \cdot a_3 \cdot \dots \cdot a_n)^{1/n} = \prod_{k=1}^{\infty} \left[1 + \frac{1}{k\cdot(k+2)}\right]^{\ln(k)/\ln(2)} \approx 2.685452001\dots$$

This shows that the limit of the geometric mean of almost all real numbers is a constant which is computed from an infinite product. The exceptions to this identity are most algebraic numbers which have definite patterns for their denominators. Yet, some transcendental numbers are also excluded, including the number e.

If we were to randomly select a real number, it would be transcendental, and its continued fraction expansion would almost certainly have a set of denominators whose geometric mean approached the same number, 2.685452001 .... This is a remarkable characterization for the real numbers.

We have really only just scratched the surface of these miraculous infinite forms. Yet, through our look at sequences, series, products, and fractions, we have been able to glimpse into the infinite and discover that such patterns are not only lovely to look at and fun to play with, but they are connected to each other in a deep and fundamental manner. However, we need not stop here, for in the coming pages we can also connect our infinite forms to that most powerful engine of mathematics: the Calculus.

9

Isaac Newton (1642–1727)

In the preceding books I have laid down the principles of philosophy; principles not philosophical but mathematical . . .
ISAAC NEWTON, *Principia*[1]

The lives of two men spanned the time from medieval superstition to the birth of modern science: Galileo Galilei and Isaac Newton. The stage was set for these two remarkable men by Nicholas Copernicus (1473–1543), a Polish astronomer and mathematician (Figure 113). Copernicus overturned the well-entrenched astronomical system of Ptolemy that maintained Earth was at the center of the universe, with the Sun, planets, and stars revolving around Earth. In his 1543 publication, *On the Revolutions of the Heavenly Spheres,* Copernicus said the Sun was at the center, that Earth rotated on its axis, and claimed Earth and other heavenly bodies revolve around the Sun, following circular paths. The only problem: almost no one listened to Copernicus.

In 1609, the German astronomer Johannes Kepler (1571–1630) published *A New Astronomy* (Figure 114). Using the astronomical observations of Tycho Brahe (1546–1601), he improved on the Copernican system by abandoning circles as the path followed by planets around the Sun. This opened the way for him to discover his three famous laws:

1. The orbits of the planets are ellipses (one of our three conic curves).
2. The area swept out by a line from the sun to the planet is equal for equal times (planets move faster when close to the sun—Figure 115).
3. The ratio of the square of the duration of a planet's orbit (length of its year) to the cube of its distance from the sun is the same for all planets.

FIGURE 113. Nicholas Copernicus (1473–1543). Photograph from Brown Brothers, Sterling, PA.

This must have been a wondrous surprise for those who accepted Copernican astronomy, for it said that, capricious, unreliable spirit was not moving the planets; instead, they move perfectly according to a natural law, and most important, that natural law was mathematical in form.

If we call the time it takes for Earth to make one complete revolution around the Sun "one year," and the distance from Sun to Earth "1.0 astronomical unit" (A.U.) and we measure a planet's period in years and distance from the Sun in A.U.s, then we can express Kepler's third law as:

FIGURE 114. Johannes Kepler (1571–1630). Photograph from Brown Brothers, Sterling, PA.

$$\frac{(\text{period of orbit})^2}{(\text{mean radius of orbit})^3} = \frac{(1 \text{ year})^2}{(1 \text{ A.U.})^3} = \frac{1^2}{1^3} = 1$$

The significance of this mathematical relationship was not wasted on the astronomers of the day, because they could use it to calculate the relative distances of the planets from the sun. First they used observations to calculate the orbital periods of the various planets. Using this information and Kepler's third law, they could easily compute how far each planet's orbit was from the sun as compared to Earth. For example, the orbital period of Mars is 687 days or approximately 1.88 years. If Kepler's third

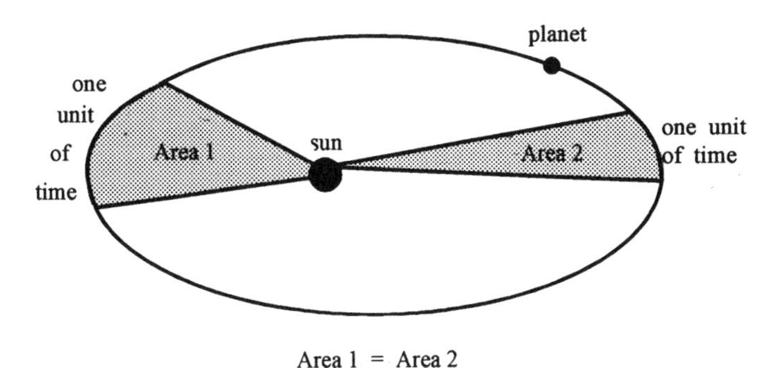

Area 1 = Area 2

FIGURE 115. Kepler's second law of planetary motion. In equal times a planet sweeps out equal areas of its elliptic orbit.

law is correct, then the square of this time divided by the cube of the mean orbital radius will equal 1, or

$$\frac{(1.88)^2}{(\text{radius of orbit})^3} = 1$$

Solving this equation we find that Mars must be approximately 1.52 A.U. or 1.52 times as far from the Sun as Earth. This, of course, is correct, since the distance from Earth to the Sun is 93 million miles and the distance from Mars to the Sun is 141.7 million miles, which is 1.52 times Earth's distance.

If the Copernican system was correct, along with Kepler's third law, then humankind might have a chance to understand the universe. Through the work of Copernicus and Kepler, the stage was set for rationalism to replace spiritualism in understanding the universe.

Galileo was born in 1564, a time when magic spirits and ancient gods were still believed to be the major forces causing worldly changes, for the vast body of people did not accept the new way of thinking. After all, the idea of supernatural forces running the universe was very ancient. The Egyptians had believed the sun was the god Ra, moving his light across the sky each day; and for the Greeks it had been Apollo. The spirits occupied matter and made it do their bidding. Spirits occupied the human body to bring about good or ill health. Spirits lived in fire and in the wind. Sometimes these spirits were beneficial to humankind, but at other times they were the source of evil. There was only one thing you could rely on in such a world—the unpredictability of the forces of nature, for there was nothing controlling the spirits or the gods.

Spanning the lives of both Galileo and Newton were the two remarkable men Rene Descartes (1596–1650) and Pierre de Fermat (1601–1665), whom we met earlier. Galileo successfully applied rationalism to account for the behavior of the universe. Descartes and Fermat built on Galileo's work, and then advanced it by attacking the tangent and area problems by inventing their wonderful coordinate system. Newton built on what Descartes and Fermat had done to finally dismiss irrational and capricious spirit as the prime mover of matter. Newton was the first to invent calculus, and with calculus he gave us the one tool we needed to understand change. This ushered in the modern age.

Isaac Newton (Figure 116) was born on Christmas Day, 1642, in the village of Woolsthorpe, England;[2] the same year that Galileo Galilei died in Italy, under house arrest by order of the pope for declaring the truth of the Copernican system. Many years later, in 1727, when Isaac Newton died, he left behind a world that was substantially different from the spirit world of Galileo's time. Certainly for many people, spirits still existed, but it was natural law, put into motion by a benevolent God that caused the wheels of the universe to turn. It was by natural law that the planets moved in perfect ellipses around the sun. By natural law universal gravity caused both the water to run down the creek bed and the moon to stay her orbit. With such wondrous natural laws, humans could hope to predict future events, for natural laws were not to be broken. The world left to us by Isaac Newton was much less scary, and much more hopeful, than the world Galileo had been born into. And in which language did these natural laws speak? Mathematics!

Of course, the old world view never completely disappeared. Even today we still hear reports of spirits suddenly appearing to meddle in the events of humans (flying saucer aliens abducting humans to conduct biological experiments and ghosts of the dead haunting a house), but their actions are small actions which are no longer required to move celestial bodies, nor to generate different species of living organisms. That is now under the control of modern science.

A CHRISTMAS CHILD

The Newtons were moderately successful farmers. They lived in the gentle river valley of Lincolnshire, England. Isaac, Newton's father, owned

FIGURE 116. Isaac Newton (1642–1727). Photograph from Brown Brothers, Sterling, PA.

a farm in the small village of Woolsthorpe, which was a half dozen miles south of the town of Grantham. None of the Newtons before Isaac Newton was literate. However, on his mother's side, Isaac's uncle, William Ayscough, had graduated with a M.A. from Cambridge University.

In October 1642, before the birth of his son, Newton's father died, leaving Newton's mother, Hannah Ayscough Newton, a pregnant widow. That Christmas Day when Isaac was born he was so small and sickly it was said he would fit into a quart mug.[3] He was not expected to live out the day. Yet, he did, developing a sufficiently strong constitution to survive into his 80s.

How young Isaac would have grown into manhood with a loving father and mother we will never know. When Isaac was only three years old his mother married a sixty-three-year-old reverend by the name of Barnabas Smith, whose church was a mere mile and a half south of Woolsthorpe in the village of North Witham. Hannah left her son in Woolsthorpe with his maternal grandmother while she moved into Barnabas Smith's church in North Witham.

Fatherless at birth, and abandoned by his mother at three, Isaac did not grow up with the love and nurturing many children take for granted. That he was not close to his Grandmother Ayscough is attested by the fact that he never mentions her with affection in any of his voluminous writings.[4] It may seem strange that a young mother would leave a three-year-old son to live fewer than two miles away. Perhaps this same thought occurred to young Newton. Under the care of his grandmother, he lived with the knowledge that his mother was only a short walk south on the road.

Abandoned and very bright, Isaac grew up in isolation. In his solitude he turned to the joy of mechanical devices, building wooden toys to amuse himself, and doll furniture for girl acquaintances. In 1653 (when Newton was eleven) Barnabas Smith died, and his mother returned to the Newton home at Woolsthorpe. One would expect a happy reunion between mother and son, but in fewer than two years Isaac was sent away to grammar school in Grantham.

His school was the Free Grammar School of King Edward VI of Grantham. When Isaac matriculated, the school was already 300 years old.[5] The primary subject of study at the Free Grammar School was Latin, including the ability to read, write, and speak the universal tongue of scholars in the seventeenth century. In Grantham, Isaac boarded at the home of Mr. Clark, who ran the local apothecary. Clark had three stepchildren by the name of Storer, two boys and a girl. Forming a dislike for the Storer brothers, Isaac developed an attachment to their sister, who was several years younger than him. A bond of friendship developed between them which, sadly, is the only suggestion of a romance in Newton's entire life. After his youth he would never develop a romantic tie to any woman.

Newton grew up with a less than developed ability to make friends with other boys. He spent his time either alone or making doll toys for Miss Storer and her friends. But he was not an ordinary boy who builds a simple toy and then abandons it after a few hours of play. Newton possessed a mind driven by curiosity in the workings of the world around him. He built a windmill run by a mouse on a treadmill. He devised a paper lantern used not only to light his way at night, but, attached to a kite, employed to scare the neighborhood's nighttime travelers.[6] While a teenager and living with the Clarks, he constructed sundials and used them to study the movement of the sun, discovering for himself the sun's solstices and equinoxes. From this experience he produced an almanac which was used by neighbors and acquaintances.

Yet, all of Newton's cleverness with his hands did not earn him friends to satisfy his social needs. In the beginning his achievements at school were ordinary, and certainly less than expected by one who would become one of the greatest mathematicians and scientists to have ever lived. Then one day a singular event occurred which seems to have turned young Newton onto a new course. While walking to school, another boy kicked Newton in the stomach. Newton challenged the attacker to a fight and fought the youth to a decisive victory.[7] Not content to win just a physical contest, Newton decided to best the vanquished in their school work. From this time on, Newton excelled in his academics.

When Isaac was seventeen, his mother brought him back to the family farm in Woolsthorpe to assist her in running the small estate. While home, Isaac fought with his mother and treated the servants badly. In less than a year she sent him back to school in Grantham. In the summer of 1661, at the age of nineteen, Isaac set off for a higher education at Trinity College, Cambridge University, the same institution attended by his uncle William Ayscough.

By Newton's time, Cambridge University was already an old, established school. During the twelfth century a number of religious orders, including the Franciscans and Dominicans, established monasteries in the town of Cambridge. By the thirteenth century these monasteries had become a university with strong religious ties to the Church of England. Within the university were separate colleges; at Newton's time they numbered sixteen, and in the present day that number has grown to thirty-one. Trinity College, Newton's school, was established in 1546. Many of England's outstanding scholars and public figures graduated from one of the Cambridge Colleges, including, of course, Isaac Newton, Oliver Cromwell, John Milton, Charles Darwin, Bertrand Russell, John Keynes, and of late, Stephen Hawking, who holds the same professorship Isaac Newton held so many years before him.

As a student at Trinity, Newton occupied the social position of a subsizar, a impoverished student who had to do menial jobs for the other students to earn his keep. Belonging to a class-conscious society in general, the subsizars were kept in their place by first serving meals in the dining hall to the other students and then enjoying their own meals from the other students' leftovers. Subsizars were required to dress differently and sit apart while in chapel. Poor Newton was not forced into this lowest class of students because his family was poor, since his mother could afford much better for her son. He was a subsizar because his mother would not provide the funds necessary to move him up into the higher ranks.

By this time Newton's personality had become etched upon his person. As described by others, he was very industrious. Indeed later he was famous for his obsession with work to the exclusion of food and rest; he was also devoutly religious, disciplined, austere, and puritanical.[8] He seems to have formed only one friendship in all his years at Cambridge—John Wickins, his roommate.

The studies at Cambridge were standard fare for a university of the time, which meant a heavy menu of Aristotle—Aristotle's logic, ethics, and rhetoric (Aristotle's syllogisms were a favorite article on exams). But Newton was not content with just the classical writings of Aristotle, for his mind was aboil with curiosity. He studied the works of Galileo Galilei, Thomas Hobbes, Robert Boyle, and most significant—the geometries of René Descartes and Pierre de Fermat. Newton was also a born experimenter. He turned part of his room at Trinity into a laboratory and used it to perform numerous experiments throughout his tenure at Cambridge. He became especially interested in light. In a daring and foolish experiment, he actually inserted a stick under his eyeball to see if the temporally deformed eye would change his vision. This was not the first time his experimentation would place him in danger. During this time Newton also became obsessed with mathematics.

During 1664 Isaac's financial position improved substantially when he was awarded a scholarship which provided him with a livery allowance and a stipend. This would guarantee his place at Cambridge for at least four more years. He graduated with his B.A. degree in 1665, but the university was closed that summer because of the spread of the plague throughout England. Newton returned to his home at Woolsthorpe to wait out the epidemic.

During the next two years, while at home, Newton achieved his best work in both mathematics and physics. While he would do much additional math and science during the balance of his life, the two great pearls which are his hallmark were worked out during the years of the plague. He invented calculus and discovered universal gravitation. Either of these achievements alone would have marked him as one of the greats in all of human intellectual endeavors. Yet, miraculously, he did both as a young man and within the span of two years. He wrote two papers on his method of fluxions, the first in 1666 and the second in 1669, which became the branch of mathematics we now call calculus. Although he did not publish these papers, he did show them to enough people to established his claim as the first to originate calculus, even though Leibniz was the first to publish his version of calculus.

In 1667 Newton returned to Cambridge and was elected a fellow of Trinity College. This meant that he would receive a wage and be given living quarters. His income, when his wage was added to his lodging, was approximately sixty pounds per year, which would compare favorably to a skilled workman who received a quarter or a third that amount, and an unskilled workman who lived on just seven to ten pounds per year.[9] In two years he was given the chair of the Lucasian professorship of mathematics that had been held by Isaac Barrow. Newton would continue to occupy this professorship until 1701, when other duties required him to permanently live in London.

At Cambridge, Newton's required duties were not taxing. Reportedly he only had to lecture for one term per year.[10] Lecturing at Trinity during the last half of the seventeenth century could, at times, require almost no work by the professors. While students were expected to attend lectures, they really received the bulk of their education from tutors, whom they paid to give them individual instruction. Hence, it was common for professors to show up to deliver a lecture on a specific day, only to find no students present! After waiting a polite time period, the professor could then retire to his private quarters. Whether this happened to Newton is unknown, but it would not have been unexpected.

With lodging and an income supplied, Newton had most of his time free to bury himself in his own studies. In his room he built a sophisticated laboratory where he could perform experiments around the clock. This included several furnaces for reducing compounds and metals. He quickly separated himself from both professors and students at Trinity, preferring to spend his time alone with his work. Even though it was expected of him, he seldom went to chapel. Most days he worked well past midnight, frequently missing meals due to his obsession with science and mathematics. In 1672 he published a paper presenting his theory on colors. To his surprise, the paper caused a number of rebuttals from established scientists. Sensitive to criticism, this experience convinced him not to push for publication in other subjects.

Because of his native intelligence and devotion to work, Newton achieved much. He discovered the general form for the binomial theorem, that rule used to expand a binomial term which has been raised to a power, i.e., $(a + b)^n$. Newton expanded the theorem to cover cases where n was a fraction and not just a positive integer. Based on this work he developed infinite series to represent the trigonometric functions $\sin(x)$ and $\cos(x)$, in addition to e^x. Always fascinated by light, he demonstrated that white light was a mixture of colors. In addition to his achievements in mathematics and gravitation, he built the first operating reflecting telescope. To do this he had to cast the mirror from a mixture of his own concoction and then grind it to the correct specifications of a parabola. The mirror was then placed in a tube and mounted on a stand. He gave his telescope to the Royal Society, after which they quickly elected him a member. He would remain a member, later acting as president, for the rest of his life.

Because of Newton's position at Trinity and his involvement with the Royal Society, others began to realize the genius behind the man. Finally, in 1687, eighteen years after he occupied the Lucasian chair in mathematics, Edmund Halley, the famed astronomer, convinced Newton to publish his theories. This book became the crowning fulfillment to his scientific career, *Philosophiae Naturalis Principia Mathematica*, or now simply known as *The Principia*. In this work Newton both detailed his work on universal gravitation and included his method of fluxions, his basic theory of calculus. It also contained the principles of theoretical mechanics (principles governing bodies in motion) and fluid dynamics (principles governing fluids in motion), the mathematical deduction of Kepler's laws of planetary motion, which explained, among other things, the mass of the Earth, Sun, and planets, the fact that Earth bulges at the equator, and the theory of tides.[11] It was an overwhelming achievement to place in just one volume. The publication took the scientific world by storm and quickly elevated Newton into the highest ranks of the scientific community.

Since we will deal with his method of fluxions and calculus in more detail it does no harm to mention here his theory of universal gravitation. In our modern world because we have so often heard about gravity, we have become jaded to the uniqueness of the idea. Aristotle, who ruled scientific thought until the Renaissance, said that something fell to earth because it was in that object's nature to do so. This idea is of no help whatsoever in understanding why or how things fall to earth. The Renaissance produced scientists who were rationalists and materialists. They believed that nature could be understood as an interlocking system of natural laws, and that the motion of objects was ultimately the result of matter in motion, bumping into other bits of matter. Hence, it was illogical to think that a force could act over a distance without any intervening matter carrying the force along. Therefore, to understand why an object fell to the ground or why Earth was attracted by the Sun required them to propose that something existed between the objects displaying the attraction. This something acted as the agent which carried the force from one object to the next.

Newton's theory of universal gravitation was diametrically opposed to such a materialistic view. He said that bodies exerted a force of attraction on each other over a distance that was directly proportional to the product of their masses, and inversely proportional to the square of their distance. In symbolic algebra this becomes: $F = Gm_1m_2/d^2$, where G is the gravitational constant, m_1 and m_2 are the two masses involved, and d^2 is the square of their distance apart. Newton went on to say that every body in the universe is affected by this law of gravitation. There is no requirement for any exotic material to flow between objects, carrying the force of gravity. Objects had mass, and mass caused gravity according to his law.

But Newton did not stop there. He proved that any law relating to force which incorporates the inverse square of the distance will result in the paths of objects being conic in nature; that is, if the force between them is inversely proportional to the square of their distance apart, then they must follow paths of parabolas, ellipses, and hyperbolas. Newton's accomplishment was an astounding result and insured that all the objects in the heavens were traveling along one of the three conic curves. Cavalieri (1598–1647), a student of Galileo's, had demonstrated that objects thrown from the surface of the earth move in parabolas, and Kepler had demonstrated that the planets move in ellipses, but it was Newton who tied it all together with his law of universal gravitation.

Newton did not restrict his studies to mathematics and science, for he possessed two other burning passions—religion and alchemy. However, both of these areas of research led him to consider ideas that could potentially bring him dishonor if known by the general public or his peers.

During this period the Church of England required belief in the Trinity—that three persons are included in the nature of God: God the Father, Christ, and the Holy Ghost. As a professor at Trinity College Newton was expected to adhere to this belief. Yet, his own tenacious study of the scriptures convinced him that the idea of the Trinity was a corruption of the scriptures during the fourth and fifth centuries. Based on his careful reading of the scriptures plus a review of the works of the early trinitarians, he maintained that God is one person, not three, and that Christ is subordinate to God. This view was called Arianism and was outlawed by the Catholic Church as early as 325 at the First Council of Nicaea. Holding such a heretical doctrine, if known, could have placed Newton's position at Trinity in great jeopardy. Therefore, he was careful to keep his deepest views on religion hidden from all but a few select confidants.

His other secret passion was the study of alchemy, which was the precursor to modern chemistry. Between the fourth and first centuries B.C., alchemy blossomed in Alexandria, Egypt. The basic goals of alchemy were to change the lesser metals, e.g., mercury, lead, and arsenic, into gold and silver. Such knowledge could also be turned toward extending life itself. The idea of alchemy grew from the Aristotelian doctrine that all things strive toward perfection. If the alchemist could find the right key he could help the base metals change into the more perfect metals. It is doubtful that Newton studied alchemy to get rich, but in light of his tremendous curiosity for all the workings of nature, he was intensely interested in seeking out the truth. He was in good company since other outstanding thinkers of the middle ages dabbled in alchemy, including Roger Bacon and St. Thomas Aquinas.

Years later when Newton died, he left handwritten manuscripts consisting of millions of words. One biographer has estimated that a million words deal with alchemy.[12] His library contained 175 books on alchemy, which represented about 10 percent of his entire library.[13] He joined London's underground society of alchemists, adopting the pseudonym of Jeova sanctus unus, which turns out to be an anagram for Isaacus Neuutonus, a latinization of his Christian name.[14] This is not the first time Newton hid his intended meaning by employing an anagram. In a letter to Leibniz he stated how his fluent and fluxion quantities addressed the fundamental theorem of calculus. He did this in such a manner that the answer would remain secret until some later date. In the letter Newton hid his statement in the anagram:[15]

6accdae13eff7i319n4o4qrr4s8t12ux

Thus, Newton's key would remain safely locked up until he chose to reveal it. This gave Newton the proof that he knew the fundamental theorem of calculus without having to divulge it. It appears that Newton's interest in religion and alchemy never left him, drawing off much of his study time in his later years.

During 1693, at the age of 51, Newton fell into a deep depression and exhaustion which lasted for a year and a half. During that time he had to curtail his study habits and take a healthier concern for his physical well-being. Little doubt exists today regarding the major cause of his illness. He carried on extensive alchemy experiments over decades which required him to handle mercury, arsenic, and other dangerous chemicals. During this evolutionary time for alchemy and chemistry it was not uncommon for Newton and others to taste their concoctions. In addition, while reducing chemicals in his furnaces, deadly fumes of heated compounds and elements would be breathed by the lungs. The symptoms displayed by Newton are virtually identical to mercury poisoning. Modern analysis of Newton's hair samples shows they contained from ten to forty times the normal concentrations of mercury.[16] After taking off a number of months and adopting a better diet, his illness abated enough for him to return to work.

In 1696 Newton left Cambridge University and moved to London, where he was appointed warden of the English mint—a substantial position. At this time his seventeen-year-old niece, Catherine Barton, moved in with him at his new London home. She was reported to be an intelligent, charming, vivacious young lady who attracted many suitors amongst the young men of London. She must have added a touch of much needed color to Newton's otherwise drab life. From his correspondence to her, it is obvious that he was exceedingly fond of her, and he signed them "Your very loving Unkle."[17]

At first, while living in London and performing his duties at the mint, he retained his position as Lucasian professor in absentia. However, in 1701 he finally gave up his Cambridge chair. This may have been motivated by the fact that he was elected to Parliament for a term during this year. However, he lost his subsequent campaign for reelection, which ended his political career.

Since 1675 Newton had been a member of the Royal Society, and in

1703 he was elected as its president, a position he would hold until his death. In recognition for his achievement at the mint (he saw England through a recoinage) and his scientific standing, Newton was knighted in 1705. Newton was now at the height of his successful career, yet he still managed to produce meaningful work. In 1704 he finished his Book *Opticks*. Then in 1707 he published an algebra book meant for the masses, *Universal Arithmetick*, which became a bestselling book.

THE LEIBNIZ INCIDENT

By 1711 it would appear that Isaac Newton was riding a crest of success and recognition that should easily carry him to retirement and eventually secure his reputation for all time. He was sixty-nine years of age and still in relatively good health. He was knighted, president of the Royal Society, and famous throughout Europe and the world. However, life is not always so accommodating. Beginning in the spring of 1711, with a letter from Leibniz to the Royal Society, a sad chain of events unfolded which demonstrates that even great men can succumb to envy and pettiness. Leibniz's letter started a bitter priority fight, not only between Newton and Leibniz, but also between the leading mathematicians of England and the Continent, over the invention of calculus.

Gottfried Wilhelm Leibniz (1646–1716) was possibly the last truly universal scholar (Figure 117). Born in Germany to a professor of moral philosophy, he taught himself Greek and Latin as a boy, entering the University of Leipzig at the age of fifteen.[18] His studies covered a broad spectrum of subjects, including law, philosophy, mathematics, logic, science, history, and theology. And he seems to have excelled at each, for like Newton, Leibniz was a genius. He completed his doctor's degree in law at the University of Altdorf in Nuremberg, Germany and then entered the country's diplomatic service.

Leibniz served in several different diplomatic posts until settling in with the Hanoverians for the balance of his long career. Diplomatic service required Leibniz to travel extensively throughout Europe, and he used this opportunity to confer with the great scientists, philosophers, and mathematicians of the day. Like Newton he had an insatiable hunger to understand everything around him. His achievements were substantial and covered so many fields they'd be difficult to track. By the age of twenty he had already published scholarly papers on both logic and law. He contin-

FIGURE 117. Gottfried Wilhelm Leibniz (1646–1716). Photograph from Brown Brothers, Sterling, PA.

ued to contribute throughout his life to the study of logic, being the first to suggest that logic was best studied using symbolic algebra—a very modern idea.[19]

His duties at the Hanover Court included serving as historian and librarian, inspiring him to write significant works on both subjects. He published accurate theories on the geology of the earth that represent one of the first attempts to explain fossils. He suggested the use of vital statistics in dealing with public health problems. He originated the science of linguistics. His works in psychology were the first to introduce the idea of the subconscious mind. In the sciences, he contributed to the idea of kinetic energy, and made many suggestions to his scientific friends who were also making significant contributions at this time. He was both engineer and architect, designing water pumps to drain silver mines, and the great formal garden at the summer palace. This all in addition to the fact he independently co-invented calculus. Leibniz left a mountain of unpublished work after his death, much of which has not been published to this day. In 1700 he founded the Berlin Academy of Science and became its first president.

There are some surprising, and other not so surprising, coincidences he shares with Newton. He also helped in a recoinage of his country, supervising the Hanover mint. He was known for his ability to concentrate with a single-minded intensity and unlock problems that had stumped others. He was clever with his hands, building a calculating machine that would not only add and subtract, but also multiply and divide. He demonstrated a version of this machine to the Royal Society on a London trip in 1673, after which he was elected to membership.

It was this London trip, and another in 1676, which helped stoke the fires of accusations and counter-accusations about who first discovered the calculus. Newton had developed his calculus between 1665 and 1666 while at home during the plague years. In October of 1666 he wrote a short paper on his calculus, but it was never published. However, in 1669 he allowed to be circulated a paper containing a fuller treatment of his method of fluxions, entitled "De analysi per aequationes numero terminorum infinitas" or "On Analysis by Infinite Series." Thus Newton earned the reputation among the English mathematicians as the creator of calculus.

During his 1673 trip to London Leibniz became interested in the state of English mathematics and asked for a summary of the subject from the current secretary of the Royal Society, Henry Oldenburg (1615–1677). After Leibniz had returned to the continent, Oldenburg sent the summary along, which contained, among other things, mention of Isaac Newton's use of infinite series to solve certain problems. During the next two years Leibniz was occupied with mathematics, working in Paris under the tutelage of the brilliant Christian Huygens (1629–1695), the mathematician who first ignited Leibniz' interest in mathematics. By 1675 Leibniz had his version of calculus worked out and he wrote a paper using the product rule for differentiation.

Leibniz visited London again in 1676 where he was shown a copy of "De analysi," containing Newton's method of fluxions. Finally, in 1684 Leibniz published his complete differential calculus in a new Leipzig journal, *Acta Eruditorum*. Then in 1686 he published, again in *Acta Eruditorum*, a paper outlining his integral calculus. In these publications Leibniz did not give credit to Newton for his fluxion method which he had clearly seen in 1676. This oversight, if it was an oversight, probably precipitated the priority war. In 1687 Newton published his own *Principia*, which contained his fluxional method.

For two decades there was no overt action. Then in 1708 an Oxford professor, John Keill, published an article on centrifugal forces in the tract of the Royal Society, *Philosophical Transactions*, where he gave Newton first credit for calculus, and mentioning Leibniz' name, refers to the *Acta Eruditorum* publication as secondary. It took a while for the Royal Society's publication to reach Leibniz, but it struck an angry chord. Leibniz sent his 1711 letter to the Royal Society asking for a retraction. Thus began a long series of publications in Europe and England, each claiming their man as the inventor of calculus and discrediting the other. It was not a battle that was waged openly. Leibniz, while publicly praising Newton, encouraged others, especially Johann Bernoulli (1667–1748), one of the famous Bernoulli family mathematicians, to write in his defense by attacking Newton. Bernoulli complied, but failed to sign his critiques, preferring to hide his own identity.

Newton was no fairer. He was president of the Royal Society, and he used this position to have his rebuttals published under the names of others. In March 1712 the Royal Society appointed an anonymous committee to investigate the two claims for priority of calculus. Early the next year they published their conclusion. Is it any surprise they gave Newton credit as the first inventor? In the footnotes they implied that Leibniz was a compulsive plagiarist.

This was too much for the Continental mathematicians. Bernoulli devised a clever trap for Newton which, he hoped, would give the advantage to his friend, Leibniz. He knew that Newton was working on a revision of his famous *Principia*, soon to be reprinted. Bernoulli had discovered an error in Proposition X of the original *Principia*, an error which he believed suggested Newton did not know the principle behind second derivatives. He decided it would be fun to publish this error after Newton's reprint was already out.

Bernoulli's plan backfired when his nephew, Nikolaus Bernoulli took a trip to London and was introduced to Newton by Abraham De Moivre (1667–1754). Nikolaus, unaware of all the skullduggery going on, innocently told Newton about his uncle's discovery. Newton was delighted, thanking both Nikolaus and Johann, and to show his appreciation, secured Johann's election as a member to the Royal Society. He, of course, corrected the error in Proposition X for the new edition of *Principia*. Only later did Newton realize that it had all been intended as a trap. In July 1713 an anonymous article known as "Charta volans" was circulated with even the name of the printer and city of origin absent. The article stated that Newton's fluxional method was clearly an imitation of Leibniz' calculus, and then went on to point out the error in Proposition X of the original *Principia* publication, all without mentioning Bernoulli's name. The article was clearly the hand of Leibniz at work to respond to the report by the Royal Society's priority committee.

The letters and counterclaims continued. Finally, in one response, Leibniz admitted to seeing some of Newton's work while in London in 1676. Newton realized that Leibniz must have been referring to his "De analysi." Before Newton could use this information, Leibniz died on November 4, 1716.

Yet, even the death of one of the claimants did not still the argument. The Continentals continued to insist that Leibniz was first, while the English tenaciously stayed loyal to their man. Modern scholars have decided from the available historical evidence that both men independently invented calculus. Newton was first to do so, but Leibniz was first to publish. It is also to the credit of Leibniz that he made a life's effort in finding the best notation for various mathematical operations, conferring with many different mathematicians. In the end, mathematicians adopted Leibniz' superior notation for calculus. In the long run the British lost, for they could not get out of the shadow of Newton and make new mathematical breakthroughs. For the next two hundred years the center of mathematical achievement would reside on the continent.

From the time of Leibniz' death, Newton lived for another eleven years. Until the last few years he was in fairly good health and continued to study and write, and remained the president of the Royal Society. On his deathbed, Newton refused to receive the sacrament of the Church of England, possibly his final protest against the idea of the Holy Trinity. He died quietly March 20, 1727.

In Newton's possessions were almost nineteen hundred books and millions of words in manuscript form. Through the efforts of his niece's husband, John Conduitt, and their daughter, Catherine Conduitt, Newton's papers were preserved, finally finding their way to the Cambridge University Library.

Newton's life was both fascinating and perplexing. He was an obvious genius, solving problems in hours which others could not solve at all. Within one of his last major publications, *The Queries*, which came out between 1704–1706, are several speculations we find hauntingly accurate in light of Einstein's theories. Query 1 asks, "Do not Bodies act upon Light at a distance, and by their action bend its Rays; . . ." In Query 30 we find, "Are not gross Bodies and Light convertible into one another?"[20] Thus Newton's ideas foreshadowed the basic tenets of special and general relativity.

Mathematics was not his main interest, physics took precedence. Mathematics was a handy tool which made it possible to solve hard problems. Unlike both Descartes and Leibniz, who enjoyed having a good time, Newton was austere, disciplined, and puritanical. In the final analysis it would have to be said that he was a loner. Yet, his gift to the rest of us was amazing in both its depth and breadth. It is only fitting that we end this review of his life with a quote from his own pen which shows his humility in the presence of truth.

> I don't know what I may seem to the world, as to myself, I seem to have been only like a boy playing on the sea shore, and diverting myself in now and then finding a smoother pebble or a prettier shell than ordinary, whilst the great ocean of truth lay all undiscovered before me.[21]

10

Calculus

Many years ago one of my good friends was married to a fascinating woman who was a disciple of occult practices. I remember her telling me of the existence of a giant cave deep within the Himalayan Mountains. She claimed inside this cave Tibetan priests had hidden ancient books filled with ancient (and secret) knowledge. I always believed the account to be anecdotal, yet it tickles the imagination to wonder what would have been in such books had they existed. If I were responsible to fill such a book with humankind's purest and most profound knowledge, what would I put inside? I don't have to think very hard. I'd fill the book with all the mathematics at my command, for mathematics is the purest, most universal, most condensed knowledge we can possess. And if I had too little room to place all my mathematical knowledge within the book, what would come first? Calculus. And we now begin a look at humankind's greatest treasure—The Calculus.

THE TANGENT

Calculus evolved from two ancient problems of trying to find the tangent to a curve and trying to find the area under a curve. First we consider the tangent problem. An obstacle for the ancient Greeks was to describe how to draw a line that would be tangent to any given curve. Before proceeding, let's consider just what it means for a line to be tangent to a curve in the classical sense. A line is tangent to a curve if it intersects the curve at *only one point without crossing the curve to the opposite side of the curve.* Try to imagine a straight line moving through two-dimensional space toward a curve in such a manner that it strikes the curve at exactly one solitary point, yet it does not cross the curve, but continues in a straight line on the same side of the curve.

Points on a curve are not set apart with gaps between them, making it easy to draw a line so that it hits just one. Every point on a curve has other points infinitesimally close to it. Because no gaps exist between points, at every small distance from any point we have infinitely many more points. Yet the tangent line somehow sneaks in and strikes just the point we have in mind, without touching a single other point, while keeping itself straight the entire time. I find the whole idea amazing. Because the points are so closely packed, for each point on a very smooth curve there exists only one line that can be its tangent. If we dare try to fit another tangent line to the same point, that second line will either miss our point; hit our point but cross over the curve to the other side; or it will strike two points on the curve. Every point on a curve has its unique tangent line. Given a point on a curve, how do we find the tangent there?

The Greeks discovered how to find the tangent line for any point on the circle. The exercise is easy. Consider the circle in Figure 118 and the point *A* on that circle. To construct the tangent to *A* we simply draw a radius from the circle's center to *A*. Once this is done, we construct a line which passes through *A* and is perpendicular to the radius. This perpen-

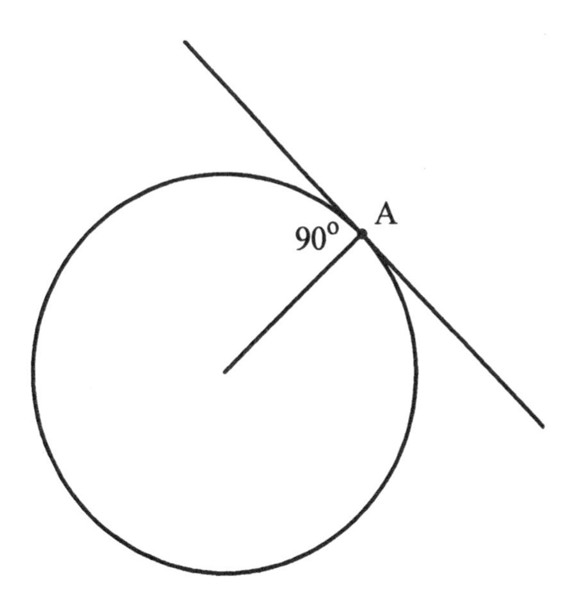

FIGURE 118. The Greeks discovered how to construct a tangent to any point on a circle. First draw a radius from the circle's center to the point. Next construct a line perpendicular to the radius at the point.

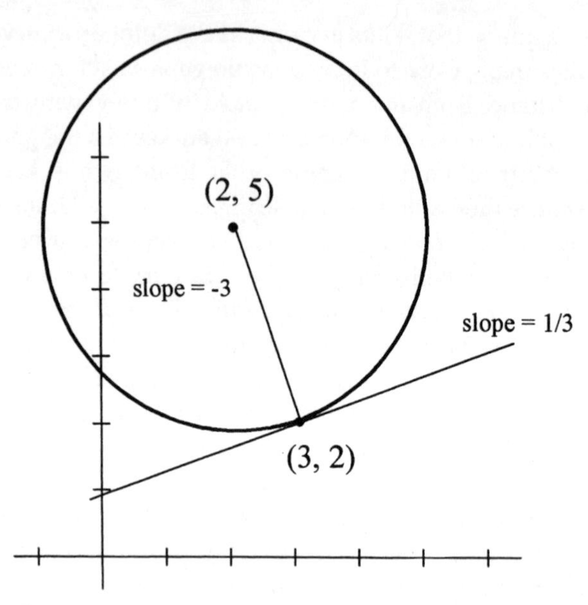

FIGURE 119. The tangent line at point (3, 2) on a circle with center at (2, 5) is $y - 2 = (1/3)(x - 3)$.

dicular line is *A*'s tangent line. If we want the actual equation for our tangent line we rely upon the modern coordinate system (which was not available to the Greeks). Suppose the center of our circle is located at the point (2, 5) as in Figure 119. Let us also suppose that our point on the circle is (3, 2). To find the tangent at (3, 2) we first find the slope for the radius connecting the center with our point. Knowing the two endpoints on the radius makes this easy; we just compute the slope according to the equation:

$$\text{slope} = \frac{\Delta y}{\Delta x} = \frac{(y_2 - y_1)}{(x_2 - x_1)}$$

where (x_1, y_1) and (x_2, y_2) are the coordinates of the two points. We can designate either point as point 1 or point 2. Choosing (3, 2) to be point 1 and (2, 5) to be point 2 we get:

$$\text{slope} = \frac{(y_2 - y_1)}{(x_2 - x_1)} = \frac{(5 - 2)}{(2 - 3)} = \frac{3}{-1} = -3$$

Hence, our slope for the radius is −3. We know that the radius is perpendicular to our tangent line, which means that the slope of the tangent line is the negative reciprocal of the slope of our radius. Hence, the tangent's slope, S, is simply ⅓. We know the slope of the tangent line and one point on that line. We use the point–slope formula to give us the equation we are after.

$$\textit{Point–slope equation: } y - y_1 = S \cdot (x - x_1)$$

where x_1 and y_1 are the coordinates of our point and S is the slope. For our example we get the equation for the tangent line:

$$y - 2 = \tfrac{1}{3}(x - 3)$$

The algebraic equation for the tangent line was beyond the range of the Greeks, but they did know how to construct the tangent line to any point on the circle by constructing a perpendicular to the radius.

Solving the tangent line problem for the circle led the Greeks to try to construct tangent lines for points on curves other than circles—on parabolas, ellipses, and hyperbolas, for example. They were mistaken if they thought it would be easy to achieve this task with the conic curves. The circle has the same curvature at every point. If you have a method of constructing a tangent line at one point, it will work at every other point. This is not the case for the conic curves, for their curvature constantly changes. Finding a method of constructing a tangent at one point may fail at another.

The Greeks did, in part, solve the tangent problem for the conics. However, they were not aware of many other curves so they made little progress on the general problem. The Greek who achieved success on the tangent problem was our man Apollonius of Perga (262–190 B.C.), who defined the conic curves in the first place. His methods are instructive for they demonstrate that the Greek approach was essentially geometric in nature as opposed to the Renaissance mathematicians, who learned to derive the algebraic expressions for the tangent lines.

For the ellipse, Apollonius drew the major axis as in Figure 120. Next he identified a point, P, on the ellipse through which he wished to draw his tangent. To do this he first dropped a line from P to hit the major axis at right angles at the point Q. The point Q now divides the major axis between the two vertices, V_1 and V_2, into the ratio V_1Q/V_2Q. Then Apollonius found a point R on the major axis lying outside the ellipse such that the ratio of RV_1 to RV_2 is the same as the previous ratio, or:

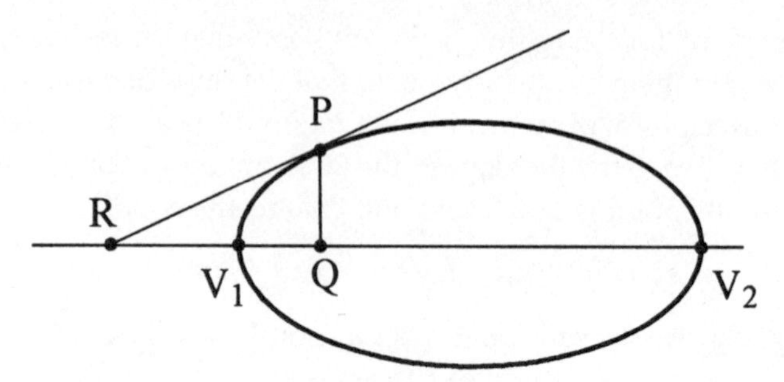

FIGURE 120. Apollonius' method of finding the tangent line at a point, P, on an ellipse with vertices at V_1 and V_2. First he dropped a perpendicular line from P to the major axis, point Q. Next he found the point R, satisfying the following proportion: $RV_1/RV_2 = QV_1/QV_2$. The line connecting R and P is his tangent.

$$\frac{RV_1}{RV_2} = \frac{QV_1}{QV_2}$$

Once the point R was found he connected the points R and P with a line, and he had his tangent line. Notice that he constructed his tangent without giving us its equation.

For the hyperbola, Apollonius found an even simpler method. First he found the asymptotic lines for the hyperbola as in Figure 121. Next he selected his point P at which he wanted to construct his tangent line. He drew a straight line so that its distance from P to the lower asymptotic line is equal to its distance to the upper asymptotic line. He proved this was the tangent line for P. In his book, *Conics*, he proved a number of additional theorems about the conic curves. As successful as Apollonius was with his techniques, he was still limited to the curves he could handle, and his methods were decidedly geometric.

Little progress was made on the general tangent problem until the Renaissance, when a number of mathematicians attacked the problem. Notably, both René Descartes and Pierre de Fermat were able to use their coordinate system to give a new twist to the problem. Descartes' method has a special charm. Consider the simple parabola $y = x^2$ in Figure 122. We select a point, P, at which we wish to find a tangent line. Descartes instructs us to find a second point, Q, also on the curve and some small distance from P. We now draw a circle that passes through both P and Q (which both lie on the curve $y = x^2$) and has its center located on the y-axis. The goal is to find the coordinates of the circle's center, for we know from the

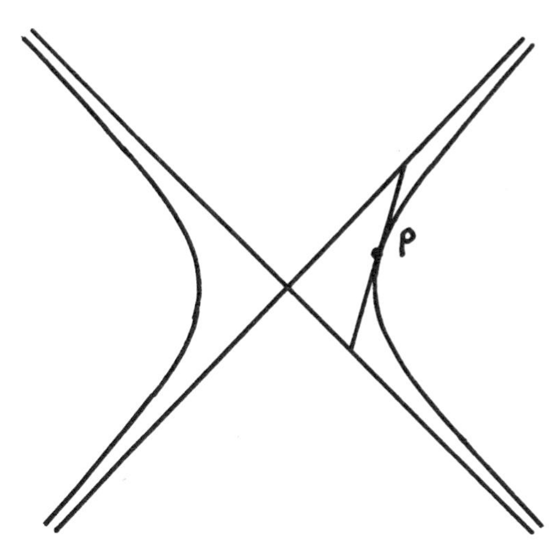

FIGURE 121. Apollonius' method of finding the tangent line to a point, P, on a hyperbola. He found the line that is bisected by P and intersects both asymptotic lines.

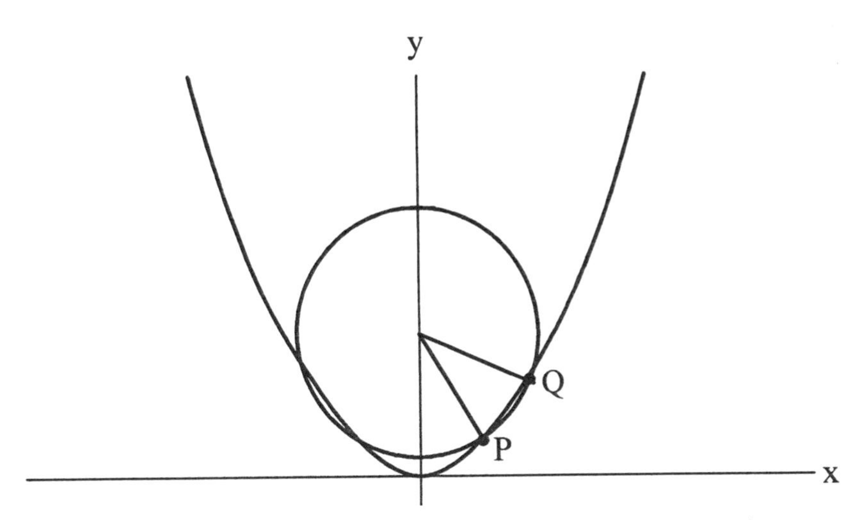

FIGURE 122. Descartes' method of finding a tangent line to the parabola $y = x^2$ at point P. He selects a second point, Q, also on the parabola. Simultaneously solving the equations of the parabola and the circle, $x^2 + (y - K)^2 = R^2$, he gets:

$$x^2 = \frac{-(1 - 2K) \pm \sqrt{(1 - 2K)^2 - 4(1)(K^2 - R^2)}}{2}$$

He throws away everything under the discriminant and solves the remainder for K giving him $K = x^2 + \frac{1}{2}$. Now, given any value for x he can find the center of the appropriate circle, and hence the tangent to the desired point.

Greeks that we can easily find a tangent line on a circle if we know the location of both the center and the point in question. Since the center is on the y-axis, the x-coordinate is simply zero, and we designate the y-coordinate as K. Therefore the location of the center of our circle is at point $(0, K)$. We must now determine the value for K.

We write the equation for the circle whose center is at $(0, K)$ and whose radius is R (which we don't know, and in fact, never know). This gives us the equation:

$$x^2 + (y - K)^2 = R^2$$

We are now going to solve the above equation and the equation $y = x^2$ simultaneously since the two points, P and Q, both lie on the circle and the parabola. Substituting x^2 for y in the circle equation we get: $x^2 + (x^2 - K)^2 = R^2$. Multiplying the terms out and collecting them we get the following fourth-degree equation in x: $x^4 + (1 - 2K)x^2 + (K^2 - R^2) = 0$. We can solve this equation for x^2 using the quadratic formula:

$$x^2 = \frac{-(1 - 2K) \pm \sqrt{(1 - 2K)^2 - 4(1)(K^2 - R^2)}}{2}$$

Now this is a complete and total mess, but Descartes tells us what to do. He says to simply throw away everything under the radical sign (the discriminant) and then solve the remainder for K. Doing just that we have: $K = x^2 + ½$. Given a specific value for x we can use it to tell us where the center of the circle is. By knowing the center of the circle, we can find the tangent line.

For example, let's find the tangent line to the parabola $y = x^2$ when x has a value of 2 (Figure 123). Substituting this into the equation for the parabola we find the coordinates of our point to be (2, 4). We now substitute $x = 2$ into the equation $K = x^2 + ½$ to get our K value. From this we discover the center of our circle to be (0, 4.5). Using the center of the circle and the point on the parabola, we find the slope of the radius connecting them is $-¼$. Taking the negative reciprocal of this value we find the slope of the tangent to be 4 and the equation for the tangent at the point (2, 4) to be $y = 4x - 4$. Thus, using Descartes' inspired circle method we have found the algebraic expression for the tangent at a point on the parabola. Descartes didn't tell us why we threw away everything under the radical sign, nor just why such a move worked. Yet, it did!

The reason it works is the following: looking at the two points in

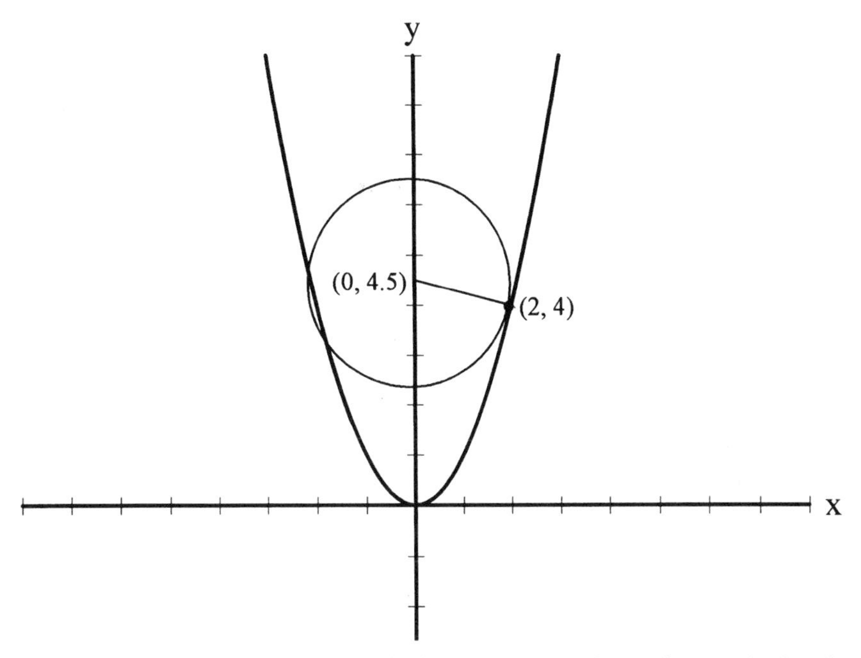

FIGURE 123. Using Descartes' method for finding the tangent line at the point (2, 4) on the parabola $y = x^2$ yields $y = 4x - 4$.

Figure 122, we can see that the tangent to the circle at P and Q will not also be tangent to the parabola. In order for this to occur, P and Q must draw closer and closer together until they are the same point. When they are the same point, then the tangent to the circle will also be the tangent to the parabola. When Descartes solved the two equations for the circle and the parabola simultaneously, he found the points where these two curves intersect, namely at P and Q. These two solutions are represented by the quadratic formula which, with the part under the radical sign (the discriminant), yields two answers (points). The only time the discriminant will be zero is when the circle intersects the parabola at only one point. Hence, when Descartes threw away the discriminant he forced the solution to be just one point. In effect, he forced the two points, P and Q, to be the same point.

Pierre de Fermat devised a method of finding tangents which evolved from his successful attempt to find the maximum or minimum value of a

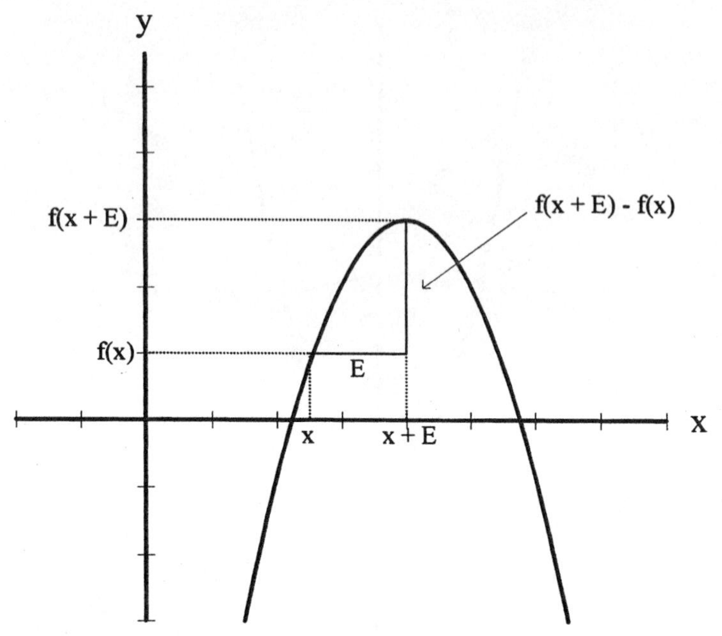

FIGURE 124. Fermat's method for finding the maximum (vertex) on the parabola, $y = -x^2 + 8x - 13$. Using Fermat's difference function:

$$\frac{f(x + E) - f(x)}{E} \approx 0$$

we find the maximum point at (4, 3).

function. Consider the parabola $y = -x^2 + 8x - 13$, whose vertex, or maximum is at point (4, 3), Figure 124. Fermat noticed that for most places on the curve of such a function, the difference between the x values of two points would not be drastically different from the difference between the y values. However, when the two points were near a maximum or minimum position of the curve, then the difference between the y values would become very small compared to the difference in the x values. Thus the ratio of the difference in the y values, expressed as $f(x + E) - f(x)$, and the difference of the x values (which he assigned the symbol E) approached zero, or:

$$\frac{f(x + E) - f(x)}{E} \approx 0$$

We can apply this approximation equation to the parabola $y = -x^2 + 8x - 13$. When we substitute $x + E$ for x we get $f(x + E) = -(x + E)^2 + 8(x + E) - 13$ and $f(x) = -x^2 + 8x - 13$. When we place these two expressions into Fermat's equation, square and multiply out the terms, and then collect like terms we end up with:

$$\frac{-2xE + 8E - E^2}{E} \approx 0$$

Dividing out the E term we have, $-2x + 8 - E \approx 0$. Now Fermat lets E be zero, realizing that when this happens at the maximum, the remaining terms should equal zero exactly, or $-2x + 8 = 0$. Solving for x we have $x = 4$, which, in fact, is the x value of the maximum of the function.

Fermat received criticism from others because his original expression had E in the denominator, and everyone knew that if E became zero, then the expression was not defined. However, the method worked so well that Fermat and others were willing to overlook this logical conundrum. Students of calculus will recognize that Fermat's approximation equation is similar to the modern derivative from differential calculus and is almost identical to the form used by Newton.

Isaac Barrow (1630–1677) was an English scholar in both ancient Greek and mathematics, who attended Trinity College, Cambridge from 1644 to 1648. Remaining at Trinity after graduation, he spent his time translating the writings of the ancient Greek mathematicians, including Euclid, Archimedes, and Apollonius. Thus, as a mathematician, he was fully aware of the tangent problem and the work attempted by the Greeks. In 1649 Oliver Cromwell won the English civil war and established the English Commonwealth. Six years later, he put down a Royalist uprising. Barrow, identified as a Loyalist, was forced to leave Cambridge. For the next five years he traveled in Europe, returning to Cambridge in 1660 when Charles II was restored to the monarchy. Barrow was appointed to both a Greek chair (professorship) and a geometry chair. This was one year before the nineteen-year-old Isaac Newton would show up at Trinity's doors. In 1663 Barrow became the first mathematician to occupy the now famous Lucasian chair of mathematics at Trinity.

While still a student at Trinity, Newton helped Barrow prepare notes for two of the professor's publications, *Lectiones Opticae* and *Lectiones Geometricae*. Thus, Newton was exposed to the tangent method as developed by Barrow. Barrow's approach was very clever. His object was to find

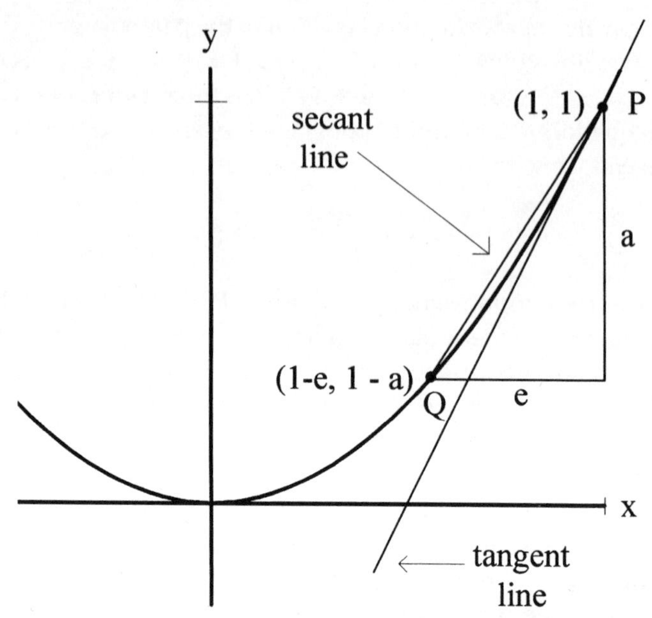

FIGURE 125. Barrow's method for finding the tangent line at point P (1, 1) on the parabola $y = x^2$. He selects a second point, Q, on $y = x^2$, and draws a secant line between P and Q. The horizontal distance between the points is designated e, and the vertical distance is a. The coordinates at Q are now $(1 - e, 1 - a)$. Barrow substitutes these values into the equation for the parabola, throws away an quadratic terms in a and e, and solves for the slope of the secant line, or $a/e = 2$. Using the slope of 2 and coordinates (1, 1) he gives the equation of the tangent line as $y - 1 = 2(x - 1)$.

the tangent line passing through a specific point on a curve. Consider again the curve in Figure 125 which is the simple parabola, $y = x^2$. Suppose we wish to find the tangent line that passes through the point (1, 1) on the parabola. If we can find the slope at this point, then we can use the point–slope equation to give us the correct tangent line. Substituting the coordinates (1, 1) into the point–slope equation we get: $y - 1 = S(x - 1)$. Now we only need to find the value of S, the slope at (1, 1).

To find S, Barrow proceeded as follows. He, like Descartes, drew a second point, Q, on the curve. He connected the two points with a line, which is called a secant line. This secant line has almost the same slope as the tangent line, but not quite. You can see at once that the closer Q is to the point (1, 1) the closer the secant line gets to the true tangent line. Next he completed the right triangle as shown in Figure 125. He defined the horizontal leg of the triangle as e, and the vertical leg as a, so that the coordinates of the point Q now become $(1 - e, 1 - a)$. Since Q is on the curve $y = x^2$, then the coordinates of Q must satisfy this equation. Barrow substitutes the coordinates of Q into $y = x^2$ to get: $(1 - a) = (1 - e)^2$. Multiplying the right side we get: $1 - a = 1 - 2e + e^2$. Now Barrow did a most peculiar thing, he just threw away the e^2 term, to leave $1 - a = 1 - 2e$. Does this sound similar to what Descartes and Fermat did?

The slope of a line is defined as the rise over the run, which in Barrow's case is simply a/e. Solving the above equation for a/e we get: $a/e = 2$. Hence the slope is 2 and the equation for the tangent line becomes $y - 1 = 2(x - 1)$, which amazingly is the exact tangent line going through the point (1, 1)! Just why this method, based on throwing the e^2 term away, worked, he wasn't sure, but work it did. Now he could find the exact tangent line for any value of x in the parabola $y = x^2$. The method worked for other curves, too. All Barrow had to do was to throw away any quadratic terms in a and e, including their product.

Mathematicians including Apollonius, Descartes, Fermat, and Barrow (and others) had now made substantial progress on the tangent problem. They could find the precise equations for tangents for specific points on all sorts of different curves. Moreover, there was still that silly little problem of throwing terms away, and just why it worked.

NEWTON'S DIFFERENTIAL

Newton's attack on the tangent problem was very similar to that used by Barrow, his teacher, and by Pierre de Fermat. However, Newton was also a brilliant physicist, and not only generalized their methods, but applied his method to a broader range of problems. For demonstration purposes, I am going to use the modern Δx in place of Newton's use of the letter o. Consider again the parabola $y = x^2$ as in Figure 126. Newton wants to know the slope of the line at point P. He begins with two points, P and Q. The difference in the x values for the two points is Δx, and the difference between the y values for the same points is just $f(x + \Delta x) - f(x)$. The slope is the ratio of these differences, or:

$$\text{slope} = \frac{f(x + \Delta x) - f(x)}{\Delta x}$$

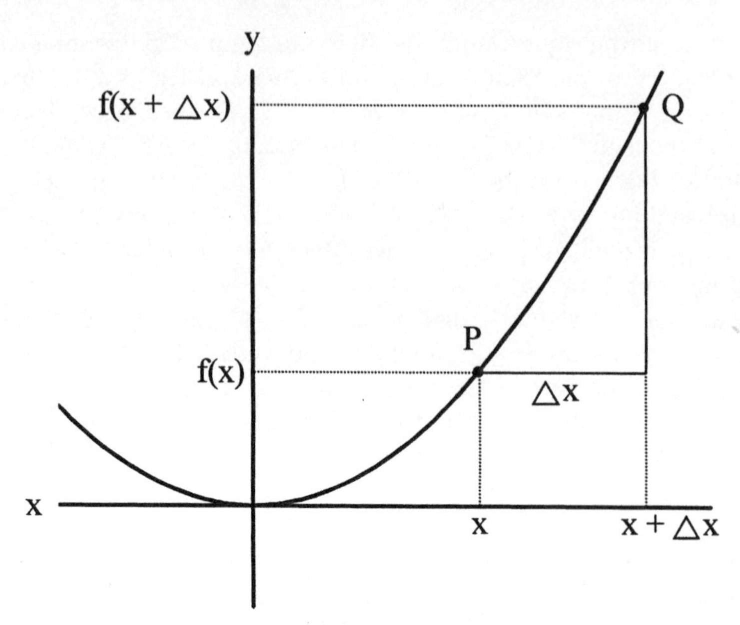

FIGURE 126. Newton's method for finding the equation for the slope of the tangent line at point P on the parabola $y = x^2$. He selects a second point Q and forms the difference equation,

$$\text{slope} = \frac{f(x + \Delta x) - f(x)}{\Delta x}$$

Newton uses the function $y = x^2$ for $f(x)$ and divides out the Δx terms. He then throws away any term still containing a Δx. The result is the equation for the slope.

The Δx (Newton used the letter o) he called an evanescent increment, and the above fraction he called his ultimate ratio of fluxions. His meaning by ultimate ratio is suggestive of our modern idea of limit. He realized that if Δx was set to exactly zero, then the slope formula would be meaningless, since we are never allowed to divide by zero. Hence, he allows Δx to be some small infinitesimal amount different from zero. As Δx gets closer to zero, then the fraction $[f(x + \Delta x) - f(x)]/\Delta x$ becomes his ultimate ratio or instantaneous change in y with respect to x.

We see at once the similarity between Newton's expression and Fermat's, for Fermat used the identical form, differing only in the choice of symbols. Now when Newton applied his difference equation to the parabola $y = x^2$ the following happened:

$$\frac{(x + \Delta x)^2 - x^2}{\Delta x} \approx \frac{x^2 + 2x \cdot \Delta x + (\Delta x)^2 - x^2}{\Delta x} \approx \frac{2x \cdot \Delta x + (\Delta x)^2}{\Delta x}$$

On the right we can just divide out the Δx to get $2x + \Delta x$. Let Δx go to zero, and we end up with the equation for the slope of the function $y = x^2$ to be the equation $y = 2x$. So far we have really not done too much differently from Newton's predecessors, except it has not been necessary to throw away any terms. However, Newton was not interested in solving for the slope at a specific point on a curve. Newton was a great organizer of knowledge, and every time he learned something new, he wanted to both organize it and extend it. Newton was after the general equation for the slope at every point satisfying the function. Newton had a grand vision enabling him to go from the particular to the universal. In our case, that is just what the equation, slope = $2x$, is. For any value on the x-axis we can substitute it into this equation and solve for the slope at that point.

Let's consider the function $y = x^3$ and see what we get when we apply Newton's difference equation.

$$\frac{(x + \Delta x)^3 - x^3}{\Delta x} \approx \frac{x^3 + 3x^2 \cdot \Delta x + 3x \cdot (\Delta x)^2 + (\Delta x)^3 - x^3}{\Delta x}$$

Again, collecting terms and dividing out the Δx, we get the equation, slope = $3x^2 + 3x \cdot \Delta x + (\Delta x)^2$. When we let Δx go to zero we end up with slope = $3x^2$. If we were to apply the method of fluxions to a fourth-order power function, $y = x^4$, we can visualize what is going to happen. The term $(x + \Delta x)^4$ will have to be expanded according to the binomial theorem. When the various terms are collected, the x^4 terms will cancel. After dividing by Δx and letting the remaining Δx terms go to zero we will be left with the term $4x^3$. We begin to see a pattern here. To find the slope of the equation x^n we simply bring the n down in front to multiply and then subtract 1 from the exponent to get $n - 1$. Hence the slope of x^n is $n \cdot x^{n-1}$. Therefore, we have a formula that gives the slope for any power function.

Even with this realization, Newton had not achieved much beyond his predecessors for there were a few mathematicians in Europe who had already stumbled onto this same formula. In expanding the expression $(x + \Delta x)^n$ we must rely upon the binomial theorem for integer values, which had been known since A.D. 1100 in China as the arithmetic triangle[1], and in the West as Pascal's triangle.

The binomial theorem when the exponent is a positive whole number is:

$$(x + y)^n =$$

$$x^n + n \cdot x^{n-1} \cdot y + \frac{n(n-1) \cdot x^{n-2} \cdot y^2}{2!} + \frac{n(n-1)(n-2) \cdot x^{n-3} \cdot y^3}{3!} + \ldots$$

This is a very nice symmetrical expression with the power of the x terms decreasing by one for each term while the power of the y term increases with each term. When we reach the $n+2$ term the $(n - n)$ expression in the numerator causes that term to be zero. Hence, for integer n, there will be exactly $n+1$ terms. Using this form of the binomial theorem it is easy to see that the expansion of $(x + \Delta x)^n$ will produce exactly $n+1$ terms, all of which will disappear under Newton's fluxion method except the term nx^{n-1}.

It is now to our advantage to look at the modern notation for Newton's fluxion method. When we apply his difference equation, we say we are taking the derivative of the original function. To do this we actually take the limit of Newton's difference equation, an idea that was not fully developed for some two hundred years after Newton.

$$f'(x) = \frac{dy}{dx} = \lim_{\Delta x \to 0} \left[\frac{f(x + \Delta x) - f(x)}{\Delta x} \right]$$

The modern notation for the derivative of the function $f(x)$ is the same $f(x)$ with a prime mark (′) following the f. The meaning of this notation is the limit of the ratio of the instantaneous change in $f(x)$ with respect to the instantaneous change in x. During and after Newton's time—but before the development of the modern concept of limits—Newton and others were criticized for using the difference equation with Δx in the denominator which became zero after manipulating the equation. However, when we use the concept of a limit, this objection is answered because we never let Δx actually become zero. We are really interested in the limit as Δx approaches zero, i.e., we are interested in a specific number which is the boundary to the process.

We have now developed the necessary tools with the derivative of a function to achieve much more than simply finding the slope of a curve. We now have the ability to analyze the phenomenon of continuous change itself. Whenever we have one variable (y) changing continuously with respect to a second variable (x) we can represent it as a derivative. Therefore we can go beyond the continuous change found in geometric curves and use the derivative to study all kinds of changing phenomena in the universe.

We have mentioned how to find the derivative of the simple power function, $y = x^n$. But there are many more complex functions we need to consider. By applying the addition principle we can find the derivative of any polynomial. The addition (and subtraction) principle simply says that if two functions are added or subtracted, we can find the derivative of the sum or difference by applying Newton's difference equations to the functions individually, or:

$$f[g(x) + h(x)]' = g'(x) + h'(x)$$

Hence, to take the derivative of a polynomial we simply find the derivative of each separate power term. For example, the derivative of the polynomial $3x^2 + 5x - 3$ becomes $6x + 5$. Taking the derivative of the -3 term we get zero. This makes sense when we consider the constant function $y = -3$ is a horizontal line passing through the y-axis at -3. Every constant function is a horizontal line, and every horizontal line has a slope of zero. Therefore the derivative of a constant function must be zero.

We can also find derivatives of transcendental functions. For example, we can apply the difference equation to the natural logarithm of x, or $\ln(x)$.

$$f'(x) = \frac{d(\ln(x))}{dx} = \lim_{\Delta x \to 0}\left[\frac{\ln(x + \Delta x) - \ln(x)}{\Delta x}\right]$$

The key now is to rewrite the right-hand limit in such a way that the Δx does not appear in the denominator. First we can use the laws of logarithms to combine the two logarithms in the numerator. At the same time we will factor out the Δx term in the denominator.

$$f'(x) = \frac{d(\ln(x))}{dx} = \lim_{\Delta x \to 0}\left[\left(\frac{1}{\Delta x}\right)\ln\left(\frac{x + \Delta x}{x}\right)\right]$$

We can now add an x term and a $1/x$ term to the left of the logarithm. At the same time, we rewrite the argument of the logarithm in the following fashion:

$$f'(x) = \frac{d(\ln(x))}{dx} = \lim_{\Delta x \to 0}\left[\left(\frac{1}{x}\right)\left(\frac{x}{\Delta x}\right)\ln\left(1 + \frac{\Delta x}{x}\right)\right]$$

Using the laws of exponents again we write the $x/\Delta x$ as an exponent of the logarithm, and we invert the $\Delta x/x$ term. We can also move the $1/x$ term outside the limit since it does not contain a Δx.

$$f'(x) = \frac{d(\ln(x))}{dx} = \left(\frac{1}{x}\right) \lim_{\Delta x \to 0} \left[\ln \left(1 + \frac{1}{\frac{x}{\Delta x}} \right)^{x/\Delta x} \right]$$

As Δx goes to zero, x divided by Δx goes to infinity. Therefore, let's replace $x/\Delta x$ with n and say that n goes to infinity.

$$f'(x) = \frac{d(\ln(x))}{dx} = \left(\frac{1}{x}\right) \lim_{\Delta x \to 0} \left[\ln \left(1 + \frac{1}{n} \right)^{n} \right]$$

The argument of the natural logarithm should now look familiar. Thus we make a most marvelous discovery, for the limit of $(1 + 1/n)^n$ as n goes to infinity is the very definition of our wonderful number e. Therefore, we can get rid of the limit symbol and substitute e for the argument of the logarithm.

$$f'(x) = \frac{d(\ln(x))}{dx} = \left(\frac{1}{x}\right) \ln(e)$$

The natural logarithm of e is just the number 1. Therefore, the final result of using the difference equation of the logarithm of x is simply:

$$f'(x) = \frac{d(\ln(x))}{dx} = \frac{1}{x}$$

It is helpful to visualize this connection between $\ln(x)$ and $1/x$. Look at Figure 127, which shows both the graph of $\ln(x)$ and $1/x$. You can see that the graph of $1/x$ agrees with the slope of $\ln(x)$. As we move to the right on the x-axis, the value of $1/x$ stays positive, but gets ever smaller. As the graph of $\ln(x)$ moves to the right, the *slope* of $\ln(x)$ gets more and more gentle as it approaches zero. The fact that the derivative of $\ln(x)$ is $1/x$ will become important when we consider the problem of areas.

Using Newton's difference equation, which really grew out of the efforts of a number of mathematicians, we can find the derivatives of all polynomials, algebraic functions, and the transcendental functions. But to get the true strength from calculus we must consider the other problem from antiquity, the area problem.

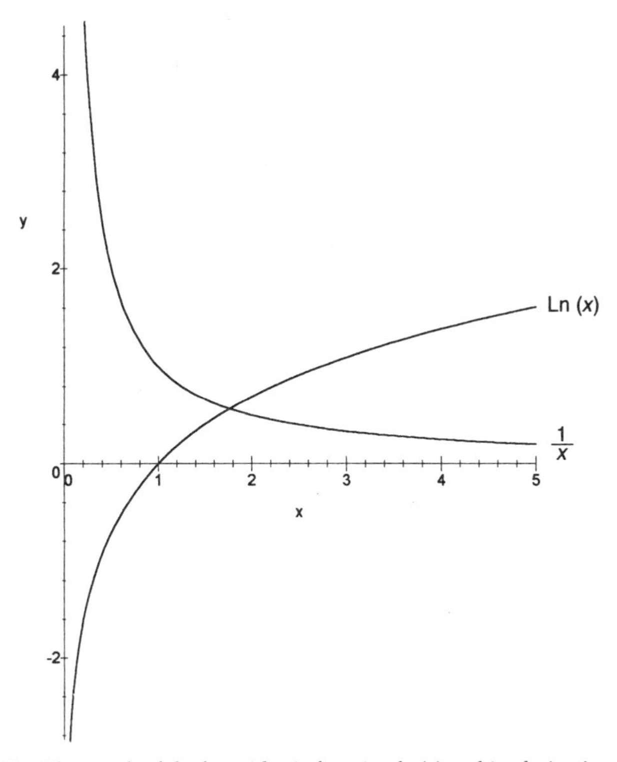

FIGURE 127. The graph of the logarithmic function $\ln(x)$ and its derivative function, $1/x$.

QUADRATURE OF AREAS

The ancient problem of quadrature, or squaring, was to find the area bounded by a curve. The first attempt was the quadrature of the circle, which they achieved using the method of exhaustion. The method of exhaustion involved filling an area bounded by a curve with one or more polygons which left some small part of the area unaccounted for. The number of polygons was then increased to decrease this unaccounted area. Hence, they were able to compute many areas and volumes to any degree of accuracy they wished. However, just being close is not always what mathematics is about, and there remained the desire to compute areas precisely, just as Hippocrates had done when he performed the quadrature of the lune.

Archimedes was a master of exhaustion, often receiving credit for the method, but he credited the method to Eudoxus who may have learned it from Hippocrates.[2] A more modern example of exhaustion was used by Kepler to show that the area of a circle is πr^2, where r is the radius of the circle. First Kepler drew a number of triangles inside of a circle as in Figure 128. If we add the area of all the triangles together, we get very close to the true area of the circle, since we are only missing that area between the base of the triangles and the circle. The more triangles we use, the closer the sum of their areas comes to that of the circle. The area of each triangle can be found by using the formula, Area = ½base × height. If we add together the areas of all n of the separate triangles in Figure 128 we get:

$$\text{Area} = \tfrac{1}{2}b_1h + \tfrac{1}{2}b_2h + \tfrac{1}{2}b_3h + \ldots + \tfrac{1}{2}b_nh$$

Kepler next factored both the ½ and h terms to get:

$$\text{Area} = \tfrac{1}{2}h(b_1 + b_2 + b_3 + \ldots + b_n)$$

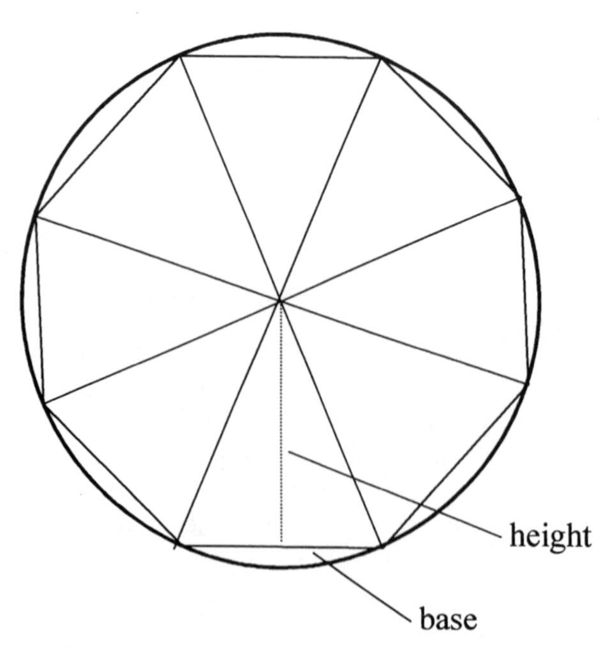

FIGURE 128. Kepler's method of finding the area to a circle. He divides the circle into a number of isosceles triangles. The sum of the area of the triangles is = $\tfrac{1}{2}b_1h + \tfrac{1}{2}b_2h + \ldots + \tfrac{1}{2}b_nh$, where b is the triangle's base and h is its height. He factors out the ½ and h to get, Area = $\tfrac{1}{2}h(b_1 + b_2 + \ldots + b_n)$. He now sets h to be the circle's radius, and the sum of the bases to be the circle's circumference; Area = $\tfrac{1}{2}r(\pi d) = \tfrac{1}{2}r(2\pi r) = \pi r^2$.

Then he realized that as the number of triangles within the circle increased, the height, h, would get closer and closer to the radius of the circle, and that the sum of the lengths of all the bases, b_i, would get closer to the circumference of the circle. Hence, he simply replaced the h term with the circle's radius and the b terms with the circle's circumference.

$$\text{Area} = \tfrac{1}{2}r(\pi d) = \tfrac{1}{2}r(2\pi r) = \pi r^2$$

Thus we have a very nice intuitive proof for the area of a circle, given that we know the circle's circumference is equal to $2\pi r$.

The next significant contribution to the quadrature problem was made by Cavalieri (1598–1647), who was a member of the order of Jesuats (as distinct from the Jesuits), a monastic order established in 1367 to help care for and bury the victims of the plague which devastated Europe during the fourteenth century.[3] Cavalieri, a student of Galileo, furthered the ideas of Kepler in regard to solving for areas under curves. He found the area of an ellipse by comparing it to a circle. Figure 129 shows both an ellipse and a circle with centers at the origin, and with the equations $x^2/a^2 + y^2/b^2 = 1$ and $x^2 + y^2 = a^2$. The distance from the origin to the intersection of both curves on the x-axis is the distance a, while the distance from the origin to the curves' intersections with the y-axis is a and b respectively. Cavalieri realized that a vertical line dropped from the circle to the x-axis would be divided by the ellipse into the ratio of b/a. Both the

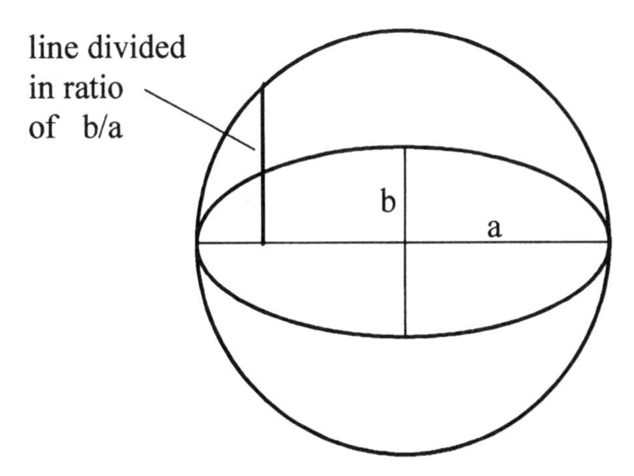

FIGURE 129. Cavalieri's method of finding the area of an ellipse. A vertical line dropped from the circle to the major axis of the ellipse is divided in the ratio of b/a where a is half the length of the major axis and b is half the length of the minor axis. Therefore, Area of Ellipse = $b/a(\pi a^2) = \pi ab$.

areas of the ellipse and the circle can be thought of as made up from an infinity of vertical lines. Hence, he reasoned, the area of the ellipse will be b/a times the area of the circle, or:

$$\text{Area of Ellipse} = b/a(\pi a^2) = \pi ab$$

Cavalieri was also able to guess that for a power curve, x^n, the area under this curve from the origin to some arbitrary distance along the x-axis, x_0, was given by $x_0^{n+1}/(n+1)$. We can see how this works in Figure 130a, where we have the straight line $y = x$. The exponent on the x is 1. Hence, the area from the origin to $x = 1$ beneath the line is just $1^2/(1 + 1) =$ ½ the area of the square, which we know is true. In Figure 130b we have the curve $y = x^2$. We want to know the fraction of the square that is under this curve. Since $n = 2$, the shaded area beneath the curve is $1^3/(2 + 1) =$ ⅓. He generalized this rule so that the area under the curve $y = x^3$ would be ¼ the area of the square, while the fraction under the curve $y = x^4$ is ⅕, and so forth.

James Gregory (1638–1675) was a Scotsman who traveled to Italy and studied the techniques of the Italian mathematicians, including Cavalieri and Torricelli. He became skilled at the method of exhaustion and determined that the area under the curve $y = 1/x$ between two points on the x-axis, a and b, is given by the equation, Area = $\ln(b) - \ln(a)$. Figure 131 shows this curve. We have shaded the area between $x = 1$ and $x = e$, our

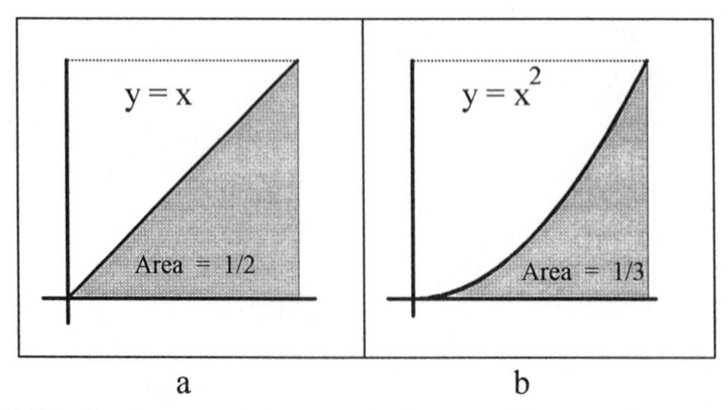

FIGURE 130. Cavalieri guessed the area under the curve of the power function x^n to be $x^{n+1}/(n+1)$. Hence, the area under the power function $y = x$ from zero to 1 is just $1^2/(1 + 1) =$ ½ and for the power function $y = x^2$ is $1^3/(2 + 1) =$ ⅓.

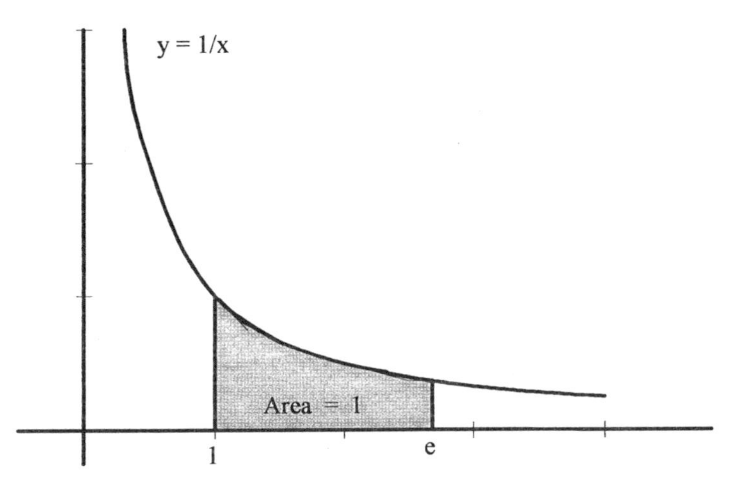

FIGURE 131. James Gregory found the area under the curve $y = 1/x$ between point a and point b to be $\ln(b) - \ln(a)$. Hence, the area between 1 and e is $\ln(e) - \ln(1) = 1 - 0 = 1$.

special constant. If we use 1 and e for our a and b, we get Area $= \ln(e) - \ln(1) = 1 - 0 = 1$. Hence, the shaded area is equal to exactly 1.

John Wallis (1616–1703) was the most talented English mathematician before Newton. Educated at Cambridge he became a professor at Oxford, and was a charter member of the Royal Society. He strived to change the nature of work on the tangent and area problems from geometry to algebra, and was strong in the techniques of infinite analysis. He knew that the ratio of areas under the power curves, $y = x^n$, to their enclosed rectangles is $1/(n + 1)$, which agrees with the work of Cavalieri. His work strongly influenced Newton.

NEWTON'S QUADRATURE

Newton noticed an amazing correspondence between the solutions to the tangent problems and the area problems. Before we investigate what this relationship is, we need to come to a fuller understanding of the final solution to the area problem, for the work up to this point has only seen the solution for specific kinds of curves, mostly power curves and conic curves. What is needed is a more generalized technique. Consider an arbitrary curve graphed in Figure 132. Under this curve we have subdivided the x-axis into three equal lengths at the points x_1, x_2, x_3, and x_4. We

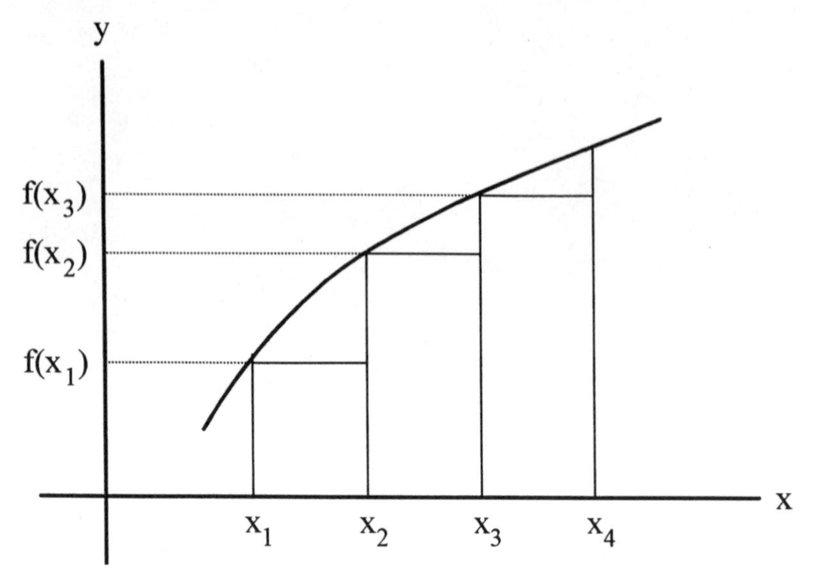

FIGURE 132. Newton's quadrature. To find the area under the general function $f(x)$ he subdivided the area under the curve into rectangles. He then summed these rectangles as the bases, Δx, went to zero.

wish to know the area under the curve between x_1 and x_4. From these subdivisions we have drawn in three rectangles. As a first approximation to the area under the curve, we can give the sum of the areas of these rectangles, or:

$$\text{Area under curve} > \text{rectangle 1} + \text{rectangle 2} + \text{rectangle 3}$$

But what is the area for these rectangles? To find it we must multiply the width of the base by the height. Since we are going to increase the number of rectangles by subdividing the bases, it is expedient to let the width of the base of each rectangle to be represented by Δx. The height of the smallest rectangle is the length of the vertical line located at x_1 from the x-axis to where the line intersects our curve. To find this value all we have to do is substitute the value of x_1 into the function that represents the curve, $f(x)$. Hence, height $= f(x_1)$. Therefore the area of the smallest rectangle is $f(x_1)\Delta x$. In like fashion the areas of the other two rectangles are $f(x_2)\Delta x$ and $f(x_3)\Delta x$. Therefore, our first approximation to the area under the curve is:

$$\text{Area under curve} > f(x_1)\Delta x + f(x_2)\Delta x + f(x_3)\Delta x$$

The area under the curve which we have missed is the sum of the three areas on top of the three rectangles. We can, of course, improve our accuracy by subdividing our x-axis at more points, giving us more rectangles. We can also write the inequality using the capital sigma.

$$\text{Area under curve} > \sum_{k=1}^{n} f(x_k) \cdot \Delta x$$

Using the idea of exhaustion we keep increasing the number of subdivisions under the curve and our computed approximate area gets closer and closer to the true area. Now we employ the modern concept of limit and ask: what is the limit to the area from all the increased subdivisions we make on the x-axis? The limit to the above expression as n goes to infinity is the area under the curve. We define the integral of the function $f(x)$ to be this limit, or:

$$\lim_{n\to\infty} \sum_{k=1}^{n} f(x_k)\Delta x = \int_a^b f(x)\,dx = \text{integral of } f(x)$$

Notice that we have replaced the limit symbol and the summation symbol (capital sigma) with the gentle curving symbol, the integral sign, which was first introduced by Leibniz. We have also replaced the Δx with the symbol dx, which stands for the instantaneous change in x. The a and b attached to the integral sign indicate the boundaries underneath the curve between which we wish to measure the area. Hence, with our integral we measure the area along the x-axis from point a to point b. Generally we include the a and b with the integral sign when we have specific boundaries in mind.

Now the question becomes, how can we easily figure out what the integral of $f(x)$ is? For specific functions, mathematicians before Newton had found specific solutions. For the power curve, $y = x^n$, they found:

$$\int x^n\,dx = \frac{x^{n+1}}{n + 1}$$

What Newton added at this point was two-fold. First, he wanted to find the areas (integrals) for more complex functions. For example, if we want to find the area bounded by the circle using integrals, which will give us an exact area and not just an approximate answer, then we must consider the curve $x^2 + y^2 = 1$. However, we need to rewrite this expression explicitly in y so that we have the form $y = f(x)$. To do this we move the x^2 term to the right and then take the square root of both sides.

$$y = \sqrt{1 - x^2} = (1 - x^2)^{1/2}$$

Technically, when we took the square root of both y^2 and $(1 - x^2)$ we should have both a + and − on the right. However, we are only going to use the + on the right so that we have a true function, and because we are going to determine the area under the first quadrant of our circle in Figure 133. We will then multiply this area by 4 to get the entire circle.

But how in the world do we figure out the integral for the function $(1 - x^2)^{1/2}$? This is where Newton made a fundamental discovery. He knew the form for the binomial theorem when the exponent is a positive integer. He discovered the general form for the binomial theorem when the exponent is negative or a fraction. The binomial theorem for positive integer values is:

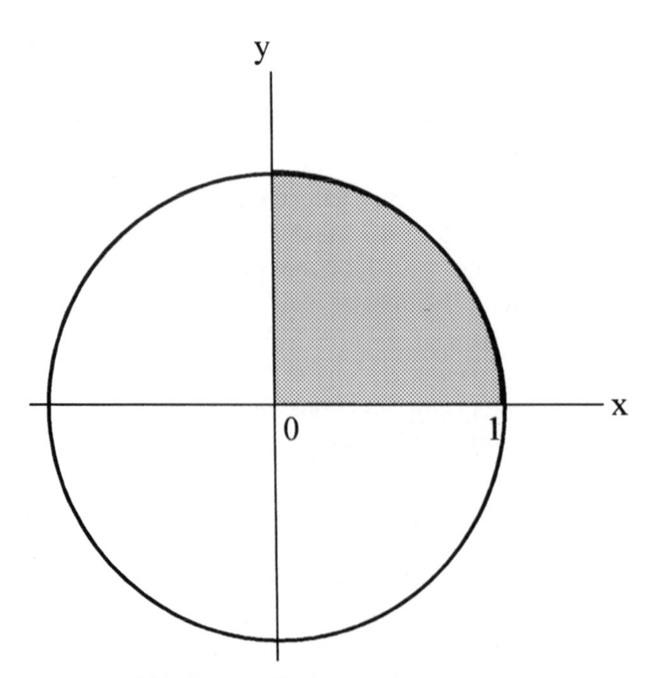

FIGURE 133. The first quadrant of a circle.

$$(a + b)^n = a^n + \frac{na^{n-1}b}{1!} + \frac{n(n-1)a^{n-2}b^2}{2!} + \frac{n(n-1)(n-2)a^{n-3}b^3}{3!} + \ldots$$

As previously mentioned, when n is a positive integer, the $n+2$ term will be zero. Hence, the expansion of $(a + b)^n$ will end after $n+1$ terms. However, Newton discovered the following form of the binomial theorem when n is the fraction m/n for the expression $(1 - x)$:

$$(1 - x)^{m/n} = 1 - \frac{m}{n}x + \frac{\left(\frac{m}{n}\right)\left(\frac{m}{n} - 1\right)}{2!}x^2 - \frac{\left(\frac{m}{n}\right)\left(\frac{m}{n} - 1\right)\left(\frac{m}{n} - 2\right)}{3!}x^3 + \ldots$$

Notice in the above expression that the signs are alternating between plus and minus, and that the numerators always have integers subtracted from the fraction m/n. This means that in none of the numerators will any of the terms ever become zero. This series is infinite, and at the same time it is the exact representation of the function on the left. In other words, Newton was able to represent the function with a fractional exponent as an infinite series in powers of x. This is not an approximation, but an infinite series which *exactly* equals our function. We already know how to find the integral of a power function, $y = x^n$. Could we use this knowledge to find the integral to an infinite series of power functions?

Newton used the above infinite series, but he substituted x^2 in place of x, and ½ in place of m/n.

$$(1 - x^2)^{1/2} = 1 - \frac{1}{2}x^2 + \frac{\frac{1}{2}\left(\frac{1}{2} - 1\right)}{2!}x^4 - \frac{\frac{1}{2}\left(\frac{1}{2} - 1\right)\left(\frac{1}{2} - 2\right)}{3!}x^6 + \ldots$$

Reducing the fractions and expanding out to six terms we get:

$$(1 - x^2)^{1/2} = 1 - \frac{1}{2}x^2 - \frac{1}{8}x^4 - \frac{1}{16}x^6 - \frac{5}{128}x^8 - \frac{7}{256}x^{10} - \ldots$$

Now that we have a representation of our function as an infinite series, we are going to try to find the integral for this function.

$$\int (1 - x^2)^{1/2}\,dx = \int \left(1 - \frac{1}{2}x^2 - \frac{1}{8}x^4 - \frac{1}{16}x^6 - \frac{5}{128}x^8 - \ldots\right)dx$$

Newton realized he could integrate the terms individually.

$$\int (1 - x^2)^{1/2}\,dx =$$

$$\int 1\,dx - \int \frac{1}{2}x^2\,dx - \int \frac{1}{8}x^4\,dx - \int \frac{1}{16}x^6\,dx - \int \frac{5}{128}x^8\,dx - \ldots$$

Using the identity: $\int x^n\,dx = x^{n+1}/(n+1)$ he got:

$$\int (1 - x^2)^{1/2} = x - \frac{\frac{1}{2}x^3}{3} - \frac{\frac{1}{8}x^5}{5} - \frac{\frac{1}{16}x^7}{7} - \frac{\frac{5}{128}x^9}{9} - \ldots$$

Which reduces simply to:

$$\int (1 - x^2)^{1/2} = x - \frac{1}{6}x^3 - \frac{1}{40}x^5 - \frac{1}{112}x^7 - \frac{5}{1152}x^9 - \frac{7}{2816}x^{11} - \ldots$$

Could the above strange infinite series really give us one-fourth the area of a circle? Consider again the circle in Figure 133, which has a radius of 1, and intersects the x-axis at (1, 0). The area of a circle is πr^2, and when $r = 1$, the area of the circle will be $\pi = 3.14159\ldots$, accurate to five places. Since we are only considering the first quadrant of our circle in Figure 133, then the value of Newton's infinite series should be $\pi/4 = 0.785398\ldots$. In order to evaluate the above infinite series for our quarter circle we must evaluate it twice where $a = 0$ and $b = 1$, because these represent the boundaries on the x-axis between which we wish to compute the area of the curve. If $s(x)$ is our infinite series then we want to compute $s(1) - s(0)$. Let's substitute $x = 1$ for b and $x = 0$ for a and subtract the results. When we substitute 0 for x, every term becomes 0, and therefore this substitution contributes nothing. Therefore, to evaluate the infinite series we must substitute 1 for x and see what we get. Substituting 1 for x in just the first six terms of the infinite series yields:

$$1 - 1/6 - 1/40 - 1/112 - 5/1152 - 7/2816 = 0.792578\ldots$$

From these six terms of our infinite series we have approximated the area to an accuracy of more than 99 percent. Newton's series converges exactly to $\pi/4$ very quickly.

Not only did Newton solve the area problem for the area of a circle with his infinite series, more significantly he introduced the idea of using infinite series containing increasing powers of x to represent functions as a legitimate procedure in mathematics. Since his time, mathematicians have become comfortable using infinite series in performing analysis, and this has greatly increased the power of doing calculus. From this we understand the deep connection between infinite series and higher mathematics.

Now for Newton's second sensational discovery, a discovery which was independently being made by Leibniz. To give you the opportunity to make the same discovery we offer two teasing clues. On the left we have the derivatives of two functions while on the right we have the integrals involving the same two functions:

$$\frac{d\left(\frac{x^{n+1}}{n+1}\right)}{dx} = x^n \qquad \int x^n\,dx = \frac{x^{n+1}}{n+1}$$

and:

$$\frac{d[\ln(x)]}{dx} = \frac{1}{x} \qquad \int \frac{1}{x}\,dx = \ln(x)$$

Do you see a correspondence between the derivative of a function and the integral of a function? Newton discovered that taking the derivative of a function is the inverse operation of finding the integral. This made it much easier to find integrals, for it is relatively easy to find derivatives but finding integrals becomes a real challenge. Many functions have derivative functions, but some of the more complex ones may not have a corresponding integral. However, once we see that differentiation and integration are the inverse operations of each other, we can use our knowledge of derivatives to help us find integrals. We can symbolically represent this wonderful discovery in the following simple expression:

$$(\int f(x))' = f(x) = \int f(x)'$$

This says that the derivative of the integral of $f(x)$ is just $f(x)$, which is the same as the integral of the derivative of $f(x)$. This correspondence between the derivative and the integral of a function is called the fundamental theorem of calculus.

To see one reason why integration (finding the integral) of a function is more difficult than finding the derivative, we begin by finding the derivative of the following polynomial: $f(x) = 3x^2 + x - 3$. Applying Newton's difference function to all three terms, $3x^2$, x, and the -3, we get:

$$d[3x^2 + x - 3]/dx = 6x + 1 - 0 = 6x + 1$$

Hence, the $3x^2$ term becomes $6x$, the x terms becomes 1, while the -3 term is 0. Now let's ask what happens when we take the derivative function, $6x + 1$, and perform integration to get back to our original $f(x)$.

$$\int (6x + 1)\,dx = \int (6x)\,dx + \int 1\,dx = 3x^2 + x$$

What happened to the -3 in the original function? When we found the derivative of $f(x)$ the -3 became 0, and when we tried to reverse the process by computing the integral, we didn't get the -3 back. We suddenly realize that when computing the integral of a function we must consider the possibility that a constant term should be added. This constant term is called the constant of integration.

THE POWER OF CALCULUS

Calculus is much more than finding tangents to curves and the areas under curves. It is really the tool needed to analyze continuous change, and continuous change is the predominate manner in which things change in our universe. The best way to demonstrate this power is through an example.

Suppose we are interested in measuring the change in the size of a population over time. We have already mentioned that populations increase according to exponential growth, but let's forget that we have discussed the issue and pretend we are beginning in complete ignorance. The population in question might be a group of humans who moved into a previously uninhabited place, or it could be the population of bacteria cells in a petri dish. Our first guess will be that the population changes at a constant rate, which is not that unreasonable of an assumption. Figure 134 shows a population that begins at a certain size, P_0, at time zero. Because the population is changing at a constant rate, we model the growth of the population with the straight line. Now let's ask what that straight line is telling us. It says that for some fixed length of time during the early stages of the population there will be a fixed magnitude of increase. It also says that at a much later time, when the population is much larger, its increase in size over the same time will be the same amount. We know this doesn't make sense. For example, a human population of 1,000 people may add 50 new members in a year. If the population grows to 1 million, do we believe it is reasonable to think it continues to increase at 50 individuals per year? Of course not.

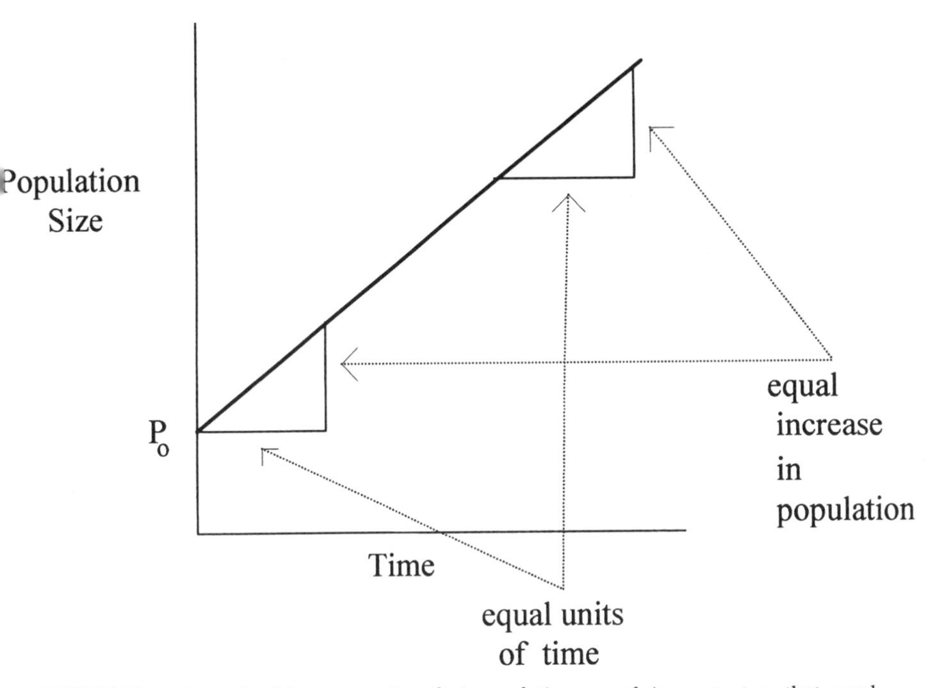

FIGURE 134. A graph of the assumption that population growth is constant, or that equal times produce equal population increases.

Instead of saying that the rate of growth is constant, let's say that the instantaneous growth is directly proportional to the size of the population. This makes sense for it says that when a population is small, the increase per unit of time is small, but when it is large, the increase for the same unit of time is large. When we make the assumption we can set the equation up using the idea of instantaneous change.

$$\frac{dP}{dt} = k \cdot P$$

The above equation can be interpreted as: the instantaneous change in the population (P) with respect to the instantaneous change in time (t) is directly proportional to the size of the population at that instant, i.e., it is equal to a constant term multiplied by the size of the population at that instant of time. This is now a reasonable assumption, for in the universe we would expect to encounter many phenomena that change proportionally to the size of the phenomena in question. The only problem is: how do we solve the above equation, which is known as a differential equation?

The first step is to get all the P and dP terms on one side of the equal sign and all the t and dt terms on the other. We do this by multiplying both sides by dt and dividing both sides by P to yield:

$$\frac{dP}{P} = k \cdot dt$$

The next step is easy, for we compute the integral of both sides:

$$\int \frac{1}{P} dP = \int k \cdot dt$$

Now we find the two integrals for the above expression. We have already mentioned that the integral for $1/x$ is $\ln(x)$. Therefore the left side becomes $\ln(P)$ plus a constant of integration. On the right side we have simply the constant function k. The integral of kdt will be just kt plus another constant of integration. Hence, we have:

$$\ln(P) + C_1 = kt + C_2$$

However, we can simplify the above expression by combining the two constants of integration into just one constant.

$$\ln(P) = k \cdot t + C$$

But we don't want the P term locked up in the argument of a logarithm, since we are after an equation which gives us the size of the population (P) as a function of time (t). We can solve this conundrum by rewriting the logarithm as an exponent. The base, of course, is the constant e because we are using the natural logarithm, and the exponent to e is $kt + C$. Hence we have:

$$P = e^{kt+C}$$

We are almost there. We can now write the expression on the right as: $P = e^C e^{kt}$. Now let's solve for the e^C term. We know that at the beginning when $t = 0$, the population is P_0 so we substitute these values in. $P_0 = e^C e^0$ or $P_0 = e^C$. Thus we discover that the constant term e^C is just the initial population. This means the final equation for our population as a function of time is simply:

$$P = P_0 e^{kt}$$

This is exactly what we had previously determined, but at that time the equation above was simply given to us. Now we have derived the proper equation using the ideas of calculus, and we can see that any instantaneous change in a phenomena in relation to time that is directly proportional to the size of the phenomena will yield The Exponential Function. This demonstrates why this function has so many applications.

In solving our problem of population growth, we actually solved a simple differential equation. Most problems in physics involve differential equations or even simultaneous systems of differential equations, and most differential equations are very difficult to solve. Currently some differential equations can be solved only with approximation methods. Entire college courses are designed to help students learn to solve certain classes of these equations, and the computer now plays a central role in this endeavor.

You can imagine Newton's excitement when he began to apply differential and integral calculus to the problems of continuous motion and was able to solve so many problems that others had struggled to understand. For instance, velocity is the speed of a moving object. It is defined as the change in distance divided by the change in time. For example, we have miles per hour (distance divided by time) as the speed of a moving car. Our instantaneous velocity is simply $v = ds/dt$, or the instantaneous change in distance (ds) divided by the instantaneous change in time (dt). This is the same as saying that velocity is the derivative of distance with respect to time.

Acceleration is the change in speed with respect to time. Hence, we have $a = dv/dt$. But we already know that $v = ds/dt$. This means that acceleration is the second derivative of distance with respect to time, or:

$$a = d[ds/dt]/dt = d^2s/dt^2$$

Newton's famous law regarding force (F) states that force is equal to mass multiplied by acceleration, $F = ma$. Using the expression above for acceleration (a), we get Newton's second law of motion.

$$F = ma = m(d^2s/dt^2) = m(dv/dt)$$

Hence, we say the force on an object is equal to its mass multiplied by the second derivative of its change in position with respect to time, or the mass multiplied by the derivative of its velocity with respect to time. Suddenly, calculus becomes the very language of physics.

The idea of instantaneous change is well accepted today, but was a foreign concept before the Renaissance. In those days to ask how fast someone traveled from one town to the next meant calculating the distance and dividing by the time to yield an overall average speed. To ask, "How fast did you go?" would be considered a strange question, because some of the time you walked your horse, and at other times you cantered. What really mattered was how long the trip required overall, and where along the road were the inns.

In our modern world of automobiles and speed limits, how fast you go at any particular moment becomes crucial as you stand before the traffic judge. The speedometers on our cars show us instantaneous velocity, and so the question, "How fast did you go?" makes perfect sense. Now the problem of the swan gliding across the pond is no longer a paradox. In the ancient world, Zeno would say that at any single moment the swan must be motionless. If its trip is made up of an infinity of motionless positions, how can it move at all? We say the swan possesses an instantaneous velocity at each and every moment, and it makes perfectly good sense because we are so used to monitoring our own speedometers to keep our speed below a certain limit. Thus we are now intuitively comfortable with the ideas of instantaneous speed in particular and instantaneous change in general. Today, all engineers and scientists are trained in the techniques of calculus so that they can use its powerful tools to uncover the deep truths hidden within instantaneous change, represented symbolically as dy/dx.

After Newton and Leibniz made their extraordinary discovery regarding the connection between differential and integral calculus, it was not until Jean Le Rond d'Alembert (1717–1783) and later Augustin–Louis Cauchy (1789–1857) successfully introduced the concept of a limit that calculus was placed on a sound theoretical basis. Today calculus is the shining jewel set in the crown of higher mathematics. For the engineer and scientist, calculus is a great and powerful tool. The prosperity we enjoy today, from electrical power, computers, powerful disease-fighting drugs, and space flight, are all the results of modern science. Yet this very science that is so productive depends upon calculus.

The mathematician feels something more toward calculus than the scientist, for mathematicians don't rely on the utility of calculus for their inspiration. To them, calculus is a glorious intellectual conquest for humanity to be enjoyed for the sheer beauty of its depths.

II

Speculations on the Nature of Mathematics

"What exactly is mathematics? Many have tried but nobody has really succeeded in defining mathematics; it is always something else."
STANISLAW ULAM (1909–1984)[1]

Our story, of course, is not complete. It cannot be, for however much we learn there is still more beyond our view. That is because the house of mathematics is so large we can never hope to visit all of its many rooms. Humans have been steadily accumulating a body of mathematical knowledge since the times of the ancient Greeks—for 2,500 years. How long have we been accumulating knowledge on the human psyche or on bacterial organisms? Most fields of study are incredibly young compared to that of mathematics. It has been known for some time that recorded mathematical knowledge is so extensive it is impossible for one person to become an expert on the entire field. If an individual can become reasonably proficient in just one small area, this is considered a substantial achievement.

The fact that the body of mathematical knowledge is immense is both bad news and good news, yet the good outweighs the bad. The bad news is that neither you nor I can grasp it all, even if we lived two or three lifetimes. However, the good news is that in our lifetimes, we will never exhaust all those beautiful equations that are waiting to charm and mystify us. How fortunate we are!

Yet, not knowing all of mathematics does not preclude us from understanding much about mathematics—that is, understanding, at least to a limited degree, what mathematics is and how it relates to the human species. We have already mentioned that mathematics is about numbers, and about other number-like objects, and that it is about relations. Yet, this definition seems to miss the mark by a wide margin, for it captures none of the subtleness or raw power of the subject. Perhaps the best way to proceed is to talk about what mathematics does for us, rather than what it is.

At the lighter end of the spectrum, mathematics entertains us. Mathematical games books are very popular, and many magazines end with some kind of "brain bogglers" or "mind benders" section which is often filled with mathematical puzzles. For the practitioner of mathematics, the daily joy of untwisting some strange mathematical relationship is always entertaining. Thus, math shares a place with other amusement activities from chess, to cards and other parlor games.

Mathematics strengthens the mind. The human brain is an organ that improves with exercise. The study of mathematics sharpens the intellect and prepares the mind for other challenging adventures. Thus it is a supplement to physical exercise, preparing the whole individual to live a full life.

Mathematics enriches with its aesthetic appeal. Moving considerably beyond the mere amusement found in games, the elegance and gracefulness of mathematical relationships touches our emotions, much like music and art can reach inside the psyche and make us feel truly alive. As an art form, mathematics can completely absorb our attention to the point of obsession. Just as great artists can be so taken with their art that they destroy their lives in its pursuit, mathematicians, too, sometimes exhibit a mania for their subject.

From reading the life histories of famous contributing mathematicians we can see a commonality. Those who contributed the most had two characteristics: they were brilliant, even genius, and they worked incredibly hard. Of course, they were unaware that it was hard work, because to them doing mathematics was as natural and as necessary as breathing itself. The greatest of the ancients, Archimedes, was known not only for his great intelligence, but for his total absorption in his contemplations. Newton, the greatest combined scientist–mathematician, was certainly a genius, but he was also compelled to work incessantly on his studies. The list goes on: Leibniz and Euler both worked with great intensity. Carl Gauss was known for his ability to concentrate on a subject until it was conquered.

Mathematics is the tool with which we unlock the productivity of technology. The citizens of industrialized nations have enjoyed an explosion in products and services during the last century. This phenomena is due to our substantial advances in the sciences, and the sciences rely on mathematics as the key that unlocks modern industrialization.

Mathematics advances the health and spread of democracy throughout the world. On the surface, this seems to be a strange connection to make. However, that form of truth which relies upon authority is secretive in nature and tends to be controlled by those in power. Authoritative truth ruled the ancient world before the Greeks. Mathematics is not a kind of secret knowledge, but by its very nature is democratic, offering itself to all who are willing to plunge into its great deductions. The pursuit of mathematics has encouraged the growth of modern rationalism, the belief that a truth is judged solely by the evidence supporting it. In turn, the triumph of rationalism has freed us to pursue new directions of thought, directions which have ultimately led to the miracles of modern medicine, space travel, and computers.

Mathematics extends our understanding of the universe. The universe seems to imitate mathematics since so many physical process are so perfectly described by mathematical constructions. We can discover mathematical relationships which are perfect descriptions of natural laws, and by using only deductive reasoning, we can manipulate these relationships until we have new expressions which suddenly reveal to us new laws of nature.

The history of mathematics and science has illustrated this process many times. When Kepler discovered that Mars' orbit around the sun was an ellipse, he did it by taking numerous sightings of the position of Mars. Hence, he discovered that his data agreed with an elliptical orbit. However, when Newton determined that planets moved around the sun in ellipses, he did it by making some basic assumptions about objects in motion and then *deducing the elliptical orbits mathematically*. He did not possess data which he matched to ellipses, but he used the inverse square law to deduce that all bodies in the universe move in one of the three conic curves. Hence, the power of his deduction was universal and demonstrated that planets moving around suns a million light years from earth also move in ellipses. What Newton discovered was a natural law that holds throughout the universe.

Einstein's discovery of his famous equation, $E = mc^2$, was the same type of mathematical discovery. He did not make this discovery based upon any experimental data. Rather, he made a basic assumption about bodies in motion. He correctly assumed that transformation equations describing bodies in motion must account for the constant speed of light. Einstein's law of motion shows that the mass of an object is affected by its velocity.

$$m = \frac{m_0}{\sqrt{1 - \frac{v^2}{c^2}}}$$

In the above equation m_0 is the resting mass of a body, v is its velocity, and c is the speed of light. If we let the velocity of the body (v) increase until it approaches the speed of light (c), then the denominator of the above equation becomes very small, making the whole term very large. Hence, according to this equation, the mass of a body increases tremendously as it approaches the speed of light. Beginning with this equation, Einstein derived the equation, $E = mc^2$. Thus mathematics gave him the power to discover a universal law of the universe. The discovery of black holes is another such example. The existence of black holes was deduced from Einstein's equations of special and general relativity. We are still in the process of verifying that they do, indeed, exist.

Since mathematics helps us discover universal laws about the universe, it increases our understanding of the universe, and is therefore a cornerstone of human knowledge.

Can we go even further? Having listed entertainment, mental exercise, aesthetics, technology, and the opportunity to understand the universe, can there be anything left? Of course! Mathematics helps us understand truths which reach beyond even the universe. A hierarchy of universality exists for truths. Some truths have almost no universality, e.g., my own name and address. Other truths enjoy more universality. Water boils at the same temperature at sea level around the world. Yet, it does not boil at this temperature on Jupiter. The truth that water at sea level boils at 100 degrees Celsius is particular to our own planet. However, Newton's and Einstein's laws of motion enjoy a much broader reach for they work everywhere in the universe.

Certain mathematical truths are the same beyond this particular universe and work for all potential universes. If we begin with the number sequence and the laws of logic, we can build an entire structure of number theory. This number theory will be the same, not only throughout this universe, but throughout every universe. Such knowledge is the most universal kind of knowledge we can possess. The universality of such mathematical insights is so compelling, that the mathematician is frequently struck dumb with wonderment.

Given the opportunity to glimpse such universal truths, can we turn our heads and refuse? No. Let us return to our fabled secret cave within the Himalayan Mountains and open the book of human wisdom. There we will record the best humans have to offer—mathematics.

ENDNOTES

INTRODUCTION

[1]Algorithm: a process or series of steps used to solve a specific problem.
[2]Randy is not his real name.

CHAPTER 1

[1]*Discover*, "Human Origins: And a One and a . . . Uh, Uh . . .," April 1997, p. 19.
[2]Ibid., June 1997, p. 14.
[3]Denise Schmandt–Besserat, *Before Writing*, Vol. 1 (Austin: University of Texas Press, 1992), p. 7.
[4]Karl Menninger, *Number Words and Number Symbols* (New York: Dover Publications, 1969), p. 11.
[5]Schmandt–Besserat, p. 188.
[6]Ibid., p. 192.
[7]David Eugene Smith, *History of Mathematics* (New York: Dover Publications, 1951), p. 43.

CHAPTER 2

[1]George F. Simmons, *Calculus Gems: Brief Lives and Memorable Mathematics* (New York: McGraw–Hill, Inc., 1992), p. 14.
[2]Kathleen Freeman, *Ancilla to The Pre-Socratic Philosophers* (Cambridge, Massachusetts: Harvard University Press, 1966), pp. 43–44.
[3]Web page, Mesopotamia Gilgamesh, www.wsu.edu8000/~dee/Mesopotamia/Gilgamesh.html.
[4]Web page, John A. Halloran, *Lexicon of Sumerian Logograms*, www.primenet.com/~seagoat/sumerian/Summerlex.html.
[5]Aristotle, *Physics* (New York: Random House, 1941), 239b, p. 335.
[6]Ibid., 204b, p. 261.
[7]Thomas Heath, *A History of Greek Mathematics* (Oxford, England: The Clarendon Press, 1960), pp. 322–331.
[8]Euclid, *The Elements*, Vol. 2 (New York: Dover, 1956), p. 395.
[9]Ibid., p. 12.
[10]Ibid., p. 142.

[11]Simmons, *Calculus Gems*, p. 35.
[12]Ibid., p. 39.

CHAPTER 3

[1]For more on this question see George Gheverghese Joseph, *The Crest of the Peacock* (London: Penguin Books, 1991), and Richard J. Gillings, *Mathematics in the Time of The Pharaohs* (New York: Dover Publications, 1972).
[2]Noah Webster, *Webster's New Twentieth Century Dictionary* (New York: Simon and Schuster, 1983), p. 1441.
[3]This is not a formal definition. A formal definition would be: a well-formed formula that can be deduced from the axioms of a formal system by the rules of logical inference.
[4]Paulo Ribenboim, *The Book of Prime Number Records* (New York: Springer–Verlag, 1989), p. 73.
[5]Carl B. Boyer, *A History of Mathematics* (New York: Princeton University Press, 1968), p. 71.
[6]Sir Thomas Heath, *A History of Greek Mathematics* (Oxford, England: The Clarendon Press, 1960), p. 182.
[7]George F. Simmons, *Calculus Gems: Brief Lives and Memorable Mathematics* (New York: McGraw–Hill, Inc., 1992) pp. 10–11.

CHAPTER 4

[1]Galileo Galilei, "Mathematics of Motion," *The World of Mathematics* (New York: Simon and Schuster, 1956), p. 734.
[2]Jane Muir, *Of Men and Numbers* (New York: Dodd, Mead & Company, 1961), p. 50.
[3]Ibid., p. 49.
[4]George F. Simmons, *Calculus Gems: Brief Lives and Memorable Mathematics* (New York: McGraw–Hill, Inc., 1992), pp. 93–94.
[5]Ibid., p. 85.
[6]Ibid., p. 96.
[7]Michael Sean Mahoney, *The Mathematical Career of Pierre de Fermat* (Princeton, New Jersey: Princeton University Press, 1973), p. 15.
[8]When I was a young mathematics student, my colleagues and I were warned by our professors to avoid trying to prove Fermat's Last Theorem, for although it is easy to state, even great mathematicians had refused to consider it, knowing it would take years from their productive careers. This warning, of course, had the opposite effect, and we all turned our hands at the problem. One night while asleep I dreamt the solution. I immediately awoke in astonishment. Rushing to the kitchen, I scribbled onto paper my supposed solution over the next several hours. In the end I had only proved a trivial and well-known identity. Such is the life of a student.

[9]An example of completing the square for the parabola,

$$8y = x^2 - 2x + 25$$

A. Move the constant to the left:

$$8y - 25 = x^2 - 2x$$

B. Add a constant to both sides that will make the right side a perfect square. To find this constant we take the coefficient to the x term (-2), divide it by 2 $(-2/2 = -1)$ and square the result $[(-1)^2 = 1]$. Hence our constant is simply 1, which we add to both sides:

$$8y - 25 + 1 = x^2 - 2x + 1$$

C. The right side is now a perfect square or:

$$8y - 24 = (x - 1)^2$$

D. We now factor the left side:

$$8(y - 3) = (x - 1)^2$$

This completes the square on the parabola.

[10]David Cohen, *Precalculus* (New York: West Publishing Company, 1997), p. 704.

CHAPTER 5

[1]*The Fibonacci Quarterly*, Fibonacci Association, % South Dakota State University Computer Science Department, Box 2201, Brookings, SD 57007-1596.

[2]The Internet address is: http://www.research.att.com/~njas/sequences/index.html.

[3]N.J.A. Sloane and S. Plouffe, *The Encyclopedia of Integer Sequences* (San Diego: Academic Press, 1995).

[4]Tobias Dantzig, *Number: The Language of Science* (New York: The Macmillan Company, 1933), p. 161.

[5]See G.H. Hardy, *A Course of Pure Mathematics* (London: Cambridge University Press, 1963); or T.J. Bromwich, *An Introduction to the Theory of Infinite Series* (New York: Chelsea Publishing Company, 1991).

[6]Carl B. Boyer, *A History of Mathematics* (New York: John Wiley and Sons, 1991), pp. 400–401.

CHAPTER 6

[1]Carl B. Boyer, *A History of Mathematics* (Princeton, New Jersey: Princeton University Press, 1968), p. 485.

[2]Florian Cajori, *A History of Mathematical Notations* (New York: Dover Publications, 1993), Sec. 552, p. 191.

[3]G.H. Hardy, *A Course of Pure Mathematics* (London: Cambridge University Press, 1963), p. 145.

[4]Such a book has been written. Eli Maor, *e: The Story of a Number* (Princeton, New Jersey: Princeton University Press, 1994).

[5]Carl Boyer, *A History of Mathematics* (New York: John Wiley and Sons, 1991), p. 35.

[6]George Cheverghese Joseph, *The Crest of the Peacock* (London: Penguin Books, 1991), pp. 115–116.
[7]Richard J. Gillings, *Mathematics in the Time of the Pharaohs* (New York: Dover Publications, 1981), p. 185.
[8]For purposes of simplification we assume a frictionless spring, which means that, once displaced, the weight will continue to bounce indefinitely.
[9]Florian Cajori, *A History of Mathematical Notations*, Sec. 498, p. 128.

CHAPTER 7

[1]Carl Boyer, *A History of Mathematics* (New York: John Wiley and Sons, 1991), p. 448.
[2]Ibid., p. 458.
[3]Eli Maor, *e: The Story of a Number* (Princeton, New Jersey: Princeton University Press, 1994), p. 123.
[4]Some may object to this characterization of Cambridge University. However, when we think of English mathematicians, we think of Wallis, Barrow, Newton, Cayley, Sylvester, Littlewood, Hardy, Russell, and Ramanujan—all Cambridge boys.
[5]Boyer, *A History of Mathematics*, p. 586.
[6]Ibid., p. 588.
[7]Ibid., p. 520.
[8]Ibid., p. 546.
[9]Lloyd Motz and Jefferson Hane Weaver, *The Story of Mathematics* (New York: Plenum Press, 1993), p. 282.
[10]Ibid., p. 285.

CHAPTER 8

[1]http://www.mathsoft.com.
[2]D.E. Smith, *History of Mathematics* (New York: Dover Publications, 1951), p. 156.
[3]Florian Cajori, *A History of Elementary Mathematics* (London: Macmillan Company, 1924), p. 150.
[4]A. Ya. Khinchin, *Continued Fractions* (New York: Dover Publications, 1992), p. 10.
[5]Ibid., pp. 86–93.

CHAPTER 9

[1]Richard S. Westfall, *Never at Rest: A Biography of Isaac Newton* (Cambridge, England: University of Cambridge Press, 1980) p. 459; *Principia*, p. 397.
[2]Newton's date of birth was December 25, 1642, according to the Julian calendar then in use. Today, in the Gregorian calendar, it would be January 4, 1643.
[3]Jane Muir, *Of Men and Numbers* (New York: Dodd, Mead & Company, 1961), p. 105.
[4]Westfall, *Never at Rest*, p. 53.

[5]Ibid., p. 55.
[6]Muir, *Of Men and Numbers*, p. 106.
[7]Westfall, *Never at Rest*, p. 60.
[8]Ibid., p. 78.
[9]Ibid., p. 181.
[10]Ibid., p. 211.
[11]George F. Simmons, *Calculus Gems* (New York: McGraw–Hill, Inc., 1992), p. 134.
[12]Westfall, *Never at Rest*, p. 290.
[13]Ibid., p. 292.
[14]This anagram works if we map the j from Jeova onto an i in Isaacus, and the v in Jeova onto a u.
[15]Newton's decoded anagram yields a statement in Latin. The English translation is, "given an equation involving any number of fluent quantities to find the fluxions, and visa versa." Westfall, *Never at Rest*, p. 265.
[16]Westfall, *Never at Rest*, p. 365.
[17]Ibid., p. 595.
[18]Carl B. Boyer, *A History of Mathematics* (New York: John Wiley and Sons, 1991), p. 400.
[19]Simmons, *Calculus Gems*, p. 148.
[20]Ibid., p. 136.
[21]Westfall, *Never at Rest*, p. 863.

CHAPTER 10

[1]Carl B. Boyer, *A History of Mathematics* (New York: John Wiley and Sons, 1991), p. 205.
[2]Ibid., p. 33.
[3]George F. Simmons, *Calculus Gems* (New York: McGraw–Hill, Inc., 1992), p. 106.

CHAPTER 11

[1]Stanislaw Ulam, *Adventures of a Mathematician* (Berkeley, California: University of California Press, 1991), pp. 273–274.

Index

Abel, Niels Henrik, 125
Analytic geometry, 4, 6, 81, 87, 89–92, 94, 113, 116, 170, 193
Apollonius of Perga, 46–52, 87, 96, 113, 251–253, 257, 259
Aquinas, Thomas, 240
Archimedes of Syracuse, 53–54, 89, 257, 266, 282
Aristarchus of Samos, 53
Aristotle, 36–37, 57–58, 64, 86, 89, 236, 238
Aryabhata the Elder, 223
Astrology, 23, 25, 229
Astronomy, 22–23, 27, 42, 46, 53, 80, 228–231
Asymptotic lines, 105, 140–141, 145, 252–253
Axiomatic theory, 69–79
Ayscough, William, 233, 235

Babylonians, 16–17, 19–21, 23, 25, 27, 31, 56, 80, 115, 157–159, 184
Bacon, Roger, 240
Barrow, Isaac, 237, 257–259
Barton, Catherine, 241
Beeckman, Isaac, 84–85
Bernoulli, Jacob, 179
Bernoulli, Johann, 245
Bernoulli, Nikolaus, 246
Binomial theorem, 238, 261–262, 272–273
Brahe, Tycho, 228
Brahmagupta, 80–81

Calculus, 4, 6, 41, 53, 74, 87, 89–90, 113–114, 126, 132, 138, 141, 144, 148, 153, 237–238, 240–241, 244–280
Calendar, 22–23, 25, 157
Cambyses II, 16, 20
Cantor, Georg, 41, 72
Cartesian coordinate system, 92–97, 99, 103, 134, 138, 176–179
Cassegrain telescope, 107–108
Cataldi, Pietro Antonio, 223
Catenary curve, 217
Cauchy, Augustin-Louis, 280
Cavalieri, 86–87, 239, 267–268
Cayley, Arthur, 186, 188–192
Chinese, 23, 80–81, 184, 261
Chuquet, Nicolas, 81
Columbus, Christopher, 52
Completeness, 78
Completing the square, 101
Consistency, 78
Constant of integration, 276, 278
Continued fraction, 83–84, 125, 221–227
Copernicus, 46, 86, 228–231
Counting, 9–14, 20
  abstract, 14–15
  concrete, 14–15, 27
Counting tokens, 13–16
Cromwell, Oliver, 235
Cusp, 138
Cyrus the Great, 16, 20

d'Alembert, Jean Le Rond, 280
Deduction, 25, 29, 44, 76, 283
De Moivre, Abraham, 246
Derivative, 262–265, 275–276
Descartes, René, 81, 84–87, 89, 91–92, 113–114, 116, 169, 176, 232, 236, 247, 252–255, 258–259
Differential equation, 278

e, 123–124, 132, 145, 147, 153, 155–157, 173–175, 181–182, 185, 205, 216–220, 224–225, 227, 264, 278
Egyptians, 16–21, 23, 25, 27, 29, 31, 43, 52, 56, 80, 115, 157–158, 231
Einstein, Albert, 209, 247, 283–284

Ellipse, 46–51, 92, 96, 98, 101–104, 106–107, 136, 145, 176–177, 180, 189–190, 228, 232, 239, 251–252, 267–268, 283
Ellipsoid, 109
Elliptic paraboloid, 111–112
Encyclopedia of Integer Sequences, 117
Eratosthenes, 52
Euclid, 42–46, 62–63, 69–70, 72, 155, 213, 257
Eudoxus, 42–44, 46, 266
Euler, Leonhard, 62, 89, 129–131, 137, 169, 175, 211, 213–214, 282

Fermat, Pierre de, 3–4, 61, 81, 84, 86–87, 89–92, 96, 113–114, 169, 176, 232, 236, 252, 255–257, 259–260
Fermat's Last Theorem, 91
Fermat's Little Theorem, 4
Fibonacci Society, 117
Fuh-hi, Emperor, 80
Functions,
  algebraic, 140, 168, 264
  elementary, 144, 168
  exponential, 146–148, 152–154, 168, 173, 205, 279
  hyperbolic cosine, 216–219
  hyperbolic sine, 216–219
  inverse, 149–151, 153
  logarithmic, 151–154,
    common, 153
    natural, 153, 155, 215, 218, 263–265, 278
  periodic, 162
  polynomial, 30–31, 138–140
  prime counting, 155
  rational, 138–141, 222
  transcendental, 140, 168, 222, 263–264
  trigonometric,
    cosine, 160, 162–168, 173–175, 180–184, 187, 204–205, 217, 219, 238
    sine, 159–168, 173–175, 180–181, 184, 187, 204–205, 217, 219, 238
    tangent, 160, 163–164
    zeta, 213, 218
Fundamental theorem of calculus, 275

Galilei, Galileo, 86–88, 182, 228, 231–232, 236, 239, 267
Gardner, Martin, 4
Gauss, Carl Friedrich, 89, 155, 169–170, 178–179, 207, 209, 282
Geometric mean, 227
Geometrical progression, 44–45
Gerstmann area, 11
Gilgamesh, 36, 116
Gödel, Kurt, 78
Goldbach's conjecture, 29
Golden mean, 32–35, 83–84, 125, 156, 224
Greeks, 22–55, 60–65, 68, 78–81, 84, 87, 96, 121, 125, 231, 248-251, 254, 257, 281, 283
Gregory, James, 130–131, 268–269

Halley, Edmund, 238
Hardy, G.H., 131
Henri's Butterfly, 184, 187
Hermann, Jacob, 179
Hindus, 81
Hipparchus, 52–53, 159
Hippasus of Metapontum, 29–30, 38
Hippocrates of Chios, 62, 64–68, 265–266
Hobbes, Thomas, 86, 236
Homo erectus, 11
Horizontal line test, 149
Hubble Telescope, 52
Huygens, Christiaan, 244
Hyperbola, 46–51, 92, 96, 98, 101, 103–108, 110–111, 145, 180, 239, 251–253
Hyperbolic Paraboloid, 111–112
Hyperboloid of one sheet, 109–110
Hyperboloid of two sheets, 110–111

Incommensurability of the diagonal, 29, 39, 43, 88
Infinite products, 211–221, 227
Integral, 271, 275–276, 278
Intuitionists, 43

Keill, John, 245
Kepler, Johannes, 155, 228–231, 238–239, 266–267, 283
Keynes, John, 235
Khinchin, Aleksandr Yakovlevich, 227
Khufu, Pharaoh, 16, 34–35

Leibniz, Gottfried, 53, 86, 114, 126, 132, 144, 237, 240, 242–247, 271, 275, 280, 282
Lemniscate, 181, 184
Leonardo of Pisa, 117
Limaçon with inner loop, 180, 182
Limit, 53, 115, 117, 122–126, 145, 156, 219–221, 227, 260, 262, 264, 271, 280

Limit point, 74–75
Lithotripter, 106
Lobachevsky, Nicolai Ivanovitch, 208
Logarithmic spiral, 181–182, 185–186

Mapping, 9, 134–136, 141, 148–150
Mathematical myths, 2–6
Matrices, 6, 184–193, 220–221
  identity, 198–199, 203
  inverse, 198–201, 203–204
  singular, 199
Mayans, 157
Menses, Pharaoh, 16
Mersenne, Father Marin, 86, 89–91
Method of exhaustion, 53, 265–266, 271
Milton, John, 235
Minkowski, Hermann, 209
Music, 1–2, 5–8, 11–12, 27–28, 82–83, 282

Napier, John, 154
Neanderthals, 10–11
Nested radical, 83–84, 125, 220
Newton, Hannah Ayscough, 233
Newton, Isaac, 53, 86, 89, 113, 144, 179, 186, 212, 232–247, 257, 259–262, 269, 271–275, 279–280, 282–284
Number line, 31, 37–42, 63, 71–75
Number theory, 3–4, 27, 43, 89–90, 114, 155, 223, 284
Numbers,
  algebraic, 31–32, 41, 63, 224–225, 227
  Bernoulli, 130
  complex, 133, 169–175, 193, 198, 205
  composite, 3, 28, 62
  counting, 10, 20, 29, 32, 71, 75, 169
  Fermat, 62
  Fibonacci, 125
  integers, 5, 20, 29, 30–32, 39, 59
  irrational, 31, 74, 145, 169
  negative, 80–81
  prime, 3, 28–29, 62, 127, 155, 213–214, 218–222
  rational, 31–32, 63, 73–74, 80, 145
  real, 30–32
  roots, 30, 144–145, 171, 173, 227
  square, 28, 59, 61
  surds, 30–32, 224
  transcendental, 32, 41, 63, 224–225, 227
  triangular, 28, 132
  unit fractions, 19

Oldenburg, Henry, 244
Oresme, Nicole, 126

Papyrus scrolls, 18, 25, 158
Parabola, 46–52, 87, 92, 96, 98–102, 104, 106–107, 111, 136–137, 145, 148–150, 176–178, 180, 216, 238–239, 251, 253–256, 258-260
Paraboloid of revolution, 111
Parmenides, 35–36
Pascal, Blaise, 114, 261
Peano, Giuseppe, 70–72
Pi, 53, 63, 73, 123, 128–132, 145–146, 156, 173–175, 212–214, 216–219, 224–225, 266–267
Plato, 6, 27–28, 42, 116
Postulates, 44, 69–70
Proof, 25–26, 29, 39, 53, 55–79, 82, 116, 126
Ptolemy, Claudius, 42, 46, 159, 228
Pyramids, 16–19, 34–35, 158–159
Pythagoras, 26–30, 38–39, 42–43, 55, 169
Pythagorean theorem, 20, 29, 68, 97, 178–179, 205–207

Quadrature of the Lune, 62–69

Ramanujan, Srinivasa, 131–132
Rational expressions, 138–140
Riemann, Georg Friedrich Bernhard, 207–210, 213
Rose with four petals, 181, 183

Schmandt-Besserat, Denise, 13–15
Secant line, 258
Sequence, 60, 71, 115–118
  converging, 117
  Fibonacci, 117, 124–125
Series, 59, 116, 118–132
  alternating, 130–131
  convergent, 126–128, 130–132
  divergent, 118, 120, 125–128
  evanescent, 127
  harmonic, 125–128, 131, 156, 214–215
  infinite, 60, 115, 147, 153–154, 168, 173–175, 183, 204–205, 211–218, 225, 238, 244, 273–275
  monotonic, 127
Set theory, 27, 71–72
Simmons, George, 79
Simultaneous equations, 194–198, 204

Smith, Barnabas, 234
Squaring the circle, 63
Sumerians, 7, 13–21, 29, 36
Syllogism, 58–59
Sylvester, James, 186, 188–189, 192

Tangent line, 248–255, 257–260, 269, 276
Thales, 23–27, 29, 56
Torricelli, Evangelista, 86, 182–183, 186, 268
Transformation, 184, 189–192, 201, 204, 209, 284

Ulam, Stanislaw, 281

Vertical line test, 136, 149–150
Viète, Francois, 212

Wallis, John, 212, 269
Whispering gallery, 106–107
Wickins, John, 236
Wiles, Andrew, 91
Writing, 13–16, 18

Zeno of Elea, 35–37, 39, 41, 280